Advance Praise for
Biodiesel Power

With unbelievable commitment, acerbic wit, and a
bona fide love for the down-home roots of the biofuel
movement, Lyle Estill has chronicled (like no other)
the complex and often entertaining dynamics of the
burgeoning world of biodiesel.

— Daryl Hannah, actress & biofuel activist
& Charris Ford, the Granola Ayatollah of Canola

It is so reassuring that good biodiesel literature
is finally beginning to emerge, with the books by
Greg Pahl, the ISU professors, and now Lyle Estill.
What most thrills me about *Biodiesel Power* is
that someone is finally telling the story of this
movement, with all its ups and downs. As a reader,
it's got me far too distracted from work right now.
As a contributor to the process that Lyle has
chronicled, I've never felt such an honor.
As therapy, this book will add years to my life.

— Kumar Plocher, president, Yokayo Biofuels, Inc.

Biodiesel Power informs and inspires, offering help and hope to the growing number of people who recognize that biodiesel can and should play a significant role in solving the world's energy problems. Estill's plainspoken language will demystify biodiesel for the biodiesel novice, while the detailed accounts of his own trials, errors, and successes will aid the home-brewer who is struggling to make consistently good fuel. The growing number of small producers will especially benefit from the author's insight into the politics surrounding the varied players in the biodiesel industry, but his story will be valuable to anyone interested in making biodiesel available and affordable.

— Emily Brewster, reference book editor
and biodiesel user

biodiesel power

biodiesel power

**The passion,
the people, and
the politics
of the next
renewable fuel**

Lyle Estill

NEW SOCIETY PUBLISHERS

Cataloging in Publication Data:
A catalog record for this publication is available from the
National Library of Canada.

Cover design by Diane MacIntosh. Image: Getty Images.

Printed in Canada.
Second printing July 2006.
Paperback ISBN 13: 978-0-86571-541-7
Paperback ISBN 10: 0-86571-541-6

Inquiries regarding requests to reprint all or part of *Biodiesel
Power* should be addressed to New Society Publishers at the
address below.

To order directly from the publishers, please call toll-free
(North America) 1-800-567-6772, or order online at
www.newsociety.com

Any other inquiries can be directed by mail to:

New Society Publishers
P.O. Box 189, Gabriola Island, BC V0R 1X0, Canada
1-800-567-6772

New Society Publishers' mission is to publish books that con-
tribute in fundamental ways to building an ecologically
sustainable and just society, and to do so with the least possible
impact on the environment, in a manner that models this
vision. We are committed to doing this not just through education, but through action. We are acting on our commitment to
the world's remaining ancient forests by phasing out our paper
supply from ancient forests worldwide. This book is one step
toward ending global deforestation and climate change. It is
printed on acid-free paper that is **100% old growth forest-free**
(100% post-consumer recycled), processed chlorine free, and
printed with vegetable-based, low-VOC inks. For further information, or to browse our full list of books and purchase
securely, visit our website at: www.newsociety.com

NEW SOCIETY PUBLISHERS   www.newsociety.com

To my brother Glen,
who has shepherded me patiently
through many endeavors,
not the least of which is renewable energy.

Contents

Acknowledgments

If I am to acknowledge anyone, I have to start with Tami and my four children, all of whom periodically wish "we could be normal." I once asked Tami to choose between a renovated kitchen and a biodiesel refinery. She rolled her eyes and said, "Build the refinery."

I should explain that the word "refinery" is a misnomer that is used throughout the biodiesel industry. We don't actually "refine" anything. What we seek is a good reaction, but the word "reactionary" never shows up in our vocabulary.

I need to acknowledge Rachel and Leif, who allowed me to join the garage band that is Piedmont Biofuels, which, remarkably, is still doing gigs much the same as we always did.

I also need to mention Tarus, who taught me how to blog, introduced me to the blogosphere and has kept up ever since. May he one day find himself driving a diesel vehicle.

And there is girl Mark, tireless warrior for biodiesel, who discovered my blog and syndicated it to the world. She brought the limelight to our efforts and has remained a genuine supporter.

I also would like to thank Anne Schwerin for letting us use her river house on Kilby Island as a place to retreat, escape and write in peace.

And, finally, thanks to the students of Energy Class and the readers of Energy Blog, who have offered insights and challenges and encouragement along the way.

Preface

I sometimes think of biodiesel as a giant, incomplete jigsaw puzzle stretched out across the kitchen table. Some people, such as Jon Van Gerpen, have been sitting at the table for a long time and have assembled a bunch of pieces.

New players, such as Oneas Mufandaedza, take a seat and say something like "I'm on feedstocks." We all can tell from the box that there is a giant field of yellow canola in the puzzle, but none of us has started on it yet. So we nod and start tossing the yellow pieces to him as we encounter them.

Girl Mark is working on backyard reactor design at one corner of the table, and images are starting to emerge clearly from her work. Joe Jobe from the National Biodiesel Board sits down. He wants to work on the legislative part. "I think I saw some lobbyist's shoes earlier," Rachel Burton remarks, looking up from the red brick of the college and scanning the loose pieces on the table.

People come and go from the project. It never ends. Just when we think we are making progress, someone else comes along, pulls up a seat and announces a desire to tackle some other part of the puzzle. We are welcoming and encouraging, and start tossing whatever pieces we know about to whoever is working on each section.

Certainly there are breakthroughs and great attachments along the way. Kumar Plocher, who has been quietly working on used analytical gear in the corner, finds his work connects to an edge that was assembled by Alex Hobbs at North Carolina State.

In this vision, we are just working away, as puzzle enthusiasts are apt to do. The chatter is idle, sometimes important, the camaraderie is evident and the puzzle slowly progresses.

I've always liked working on puzzles.

Introduction

B iodiesel has buzz. In the commercial fuel sector it has buzz as a lubricity additive to petroleum diesel. In the clean air crowd it has buzz for its reduced emissions. In agricultural circles it is talked about as a new cash crop. Academia is excited because biodiesel is a frontier, full of unknowns. There is plenty of groundbreaking research still to be done on how to make fuel from soy or algae or flies that feed on hog waste. Biodiesel has buzz with the peaceniks because there is "No War Required" to obtain it. It has traction with those on the right side of the political spectrum because it can be "Made in America."

And where biodiesel has the greatest buzz of all is with the early-adopters, who are running around on 100 percent biodiesel, or B100, as it is known. This small community of consumers is narrow and intense and creates tremendous word-of-mouth interest in the fuel. I have inhabited the B100 world for the past three years, making my own fuel, designing and building biodiesel reactors, experimenting with different feedstocks, catalysts and reactants and contributing to the rapidly growing grassroots movement that is biodiesel.

This is the story of Piedmont Biofuels, a small biodiesel co-op that has risen from the classroom to the backyard to commercial production. Our story is inexorably tied to the stories of others and to the fledgling biodiesel industry in the United States.

All around the edges of this book, and at its heart, are entries from my weblog, known simply as Energy Blog. It began as an innocent attempt to communicate with my night school students at our local community college and accidentally became a chronicle of the biodiesel movement.

I need to tip my hat to my fellow denizens of the lower blogosphere (those of us with fewer than a thousand discrete readers per night) because Energy Blog never really fit the mold. While others were introducing emoticons and spinning off deeply personal reflections backed by whatever music was playing at the time, I was publishing essays on BTUs.

But blogs are not books. People read blog entries when they are at work — usually when they are supposed to be doing something else. People read books on airplanes and in bed — usually on their own time. With this difference in mind I have organized this as a book supported by relevant blog entries along the way. Where necessary I have modified the original blog entries. The book is part journalism, part storytelling and part speculation on an industry that is not yet fully birthed.

Thomas Paine was the penman of the American Revolution. Ann Landers was a popular advice columnist when I was growing up. My guess is that I am somewhere between the two.

one

Stinky Kitchen

I knew the smell from the front sidewalk. It wasn't the mimosa trees. Their fragrant blooms had ended. It wasn't Ruby's gardenia beside the front porch. It had lost its flowers long ago. What I smelled as I approached the house was rancid fryer oil.

Tami had just renovated the kitchen in our dilapidated farmhouse. She and a cast of contractors had moved it from something you would find in a shack to something you would see in an upscale magazine. She was in the afterglow of a summer's work, which is why the smell of rancid grease emanating from the new kitchen caused both my pace and my heart to quicken.

When I opened the front door, I was engulfed in the putrid, wafting, solid smell of spoiled vegetable oil. Someone had used our new kitchen as a laboratory.

Someone was having a hard time making fuel down at Summer Shop and had trudged up to the house to use the stove.

Tami does not make biodiesel. At the time she didn't even drive a diesel vehicle. Her preference was an Isuzu SUV. But she had humored our quest for fuel. She had watched young idealists pass through our place. She had left us alone at the kitchen table to speculate long into many nights on how we could make biodiesel out of waste vegetable oil. And we had shown our gratitude by boiling veggie in her new kitchen.

I walked into the house and wondered if it was over. Not my marriage to Tami, but our quest for fuel in her domain.

My journey into biodiesel had begun almost two years earlier at the 23rd Annual Festival for the Eno. I boarded the exhibitors' bus and listened to the tour guide give her pitch. The bus was powered by biodiesel, which was made from soybeans, which were grown in America. That was cool. I thought about the 1962 Romanian-made tractor my father-in-law had lent me 12 years earlier. "What do I have to do to my engine to run on biodiesel?"

"Nothing," she said. "If you have half a tank of petroleum diesel and you add half a tank of biodiesel, the two instantly blend and your engine is none the wiser."

I was intrigued. "Will it gum up my fuel lines?" I asked.

"Nope," she replied cheerily. "It has a cleansing effect on the motor."

Another fellow asked what sort of mileage he would get running on biodiesel and was told the miles per gallon were the same as for normal diesel. She set a hook in my imagination. "Where can I get some?" I asked.

At this point her countenance changed. "You can't," she said. "It's available only on state contract."

I was confident she was wrong about that and I skipped off the bus, filled with the idea of running my nasty old diesel tractor on this amazing new substance. I stopped by the American Tobacco Trail booth. The fellow there thought biodiesel sounded like a great idea and he agreed that someone needed to be selling the stuff in Chatham County. He even wanted to buy some for his tractor. I stopped by the North Carolina Sustainable Energy Association booth and they too were easily excited about the fuel. They figured there surely were some tax credits available for investments in biodiesel.

After three scorching days at the Festival for the Eno, I was convinced that I should open a biodiesel gas station at my metal sculpture studio in Moncure. After all, it is an abandoned gas station on a somewhat busy road. I had visions of filling up happy customers with a renewable fuel made from vegetable oil from crops grown by happy farmers in America.

I was astonished to learn that the tour guide was correct. Three summers ago you could not buy biodiesel in North Carolina. I was disheartened by the news, and after fruitless Internet searches I realized that if I wanted to put biodiesel in my tractor I would have to make it myself.

Like so many others in this country, my entry into the world of biodiesel was through Joshua Tickell's self-published book, *From the Fryer to the Fuel Tank.* I ordered a copy off the net and read it from cover to cover.

Chemistry is not my strong suit. I spent most of my time in high school chemistry figuring out ways to get Martha Peterson to notice me. I had more luck with Martha than I did with chemical reactions, but after two brief semesters I ended up without a thorough understanding of either one.

And I have grown hesitant about new ideas. My daughter Jessalyn keeps an imaginary ledger of my "flops." When she comes to visit she is quick to say things like, "Dad, do you realize that every time I come we remove a couple of truckloads of your bad ideas from the farm?"

She is right about that. There was the failed attempt at vermiculture composting in abandoned refrigerators. She also helped me tote away the cold frame made from orphaned windows. It had rotted away before I mastered winter vegetable production. And she helped me remove piles of recycled bricks that I had accumulated in the woods.

"The snail ranch, Dad. Remember that one? You were going to raise them on garlic and cilantro so they would be pre-flavored. That's going in the "flop" column, Dad."

I knew that if I told Jess I was planning to make fuel out of waste vegetable oil she would not be moved, so instead I tried out the idea on my friend Gary.

Gary has a huge brain and he is patient with me. It is not unusual to find us, on any given Saturday or Sunday in the summer, floating on inner tubes down at my pond, surrounded by splashing children and discussing the next big idea.

He rejected the crawdad farm. He showed little interest in pawpaws. As a great conserver of personal energy, he found no merit in my establishment of a popcorn operation that would be staffed by us and our children. But Gary liked biodiesel. And he figured that, if we had a recipe, we could make some ourselves. My guess is that he paid attention in chemistry class.

People seldom dwell on my good ideas, but I have had a few along the way. I once heard a chef from Louisiana describe on the radio the taste of a deep-fried turkey and I was convinced of the absolute need to prepare one myself.

Deep-frying a turkey is a somewhat harrowing affair that involves lowering a whole bird into four or five gallons of hot oil in a pot atop a propane burner. It requires considerable attentiveness, but in the end it is not that hard. The secret is to not let the oil temperature creep past 400 degrees Fahrenheit or the oil will start to smoke and add a nasty flavor to the meat.

I can't say I have perfected the technique, but I have deep-fried a number of birds and some of them have been exquisite. Prior to my fixation with biodiesel, I would pour the waste vegetable oil out in the woods. On those occasions I noticed that the oil killed any vegetation it blanketed. I also observed it was rapidly consumed by dogs, possums, raccoons, foxes, bees, gnats and other residents of our surrounding forest. So before boarding that fateful bus to the Eno I already had a vague sense that there was good energy left in used vegetable oil.

Gary and his family came over for a deep-fried turkey feast, and for the first time I was less interested in the outcome of the meat than in our first attempt to make fuel from the leftover oil.

We retreated to a spool table at Summer Shop, leaving the children and wives behind. Gary had found a blender at the thrift shop. I had lined up some rubbing alcohol and some glassware. We had a plastic bottle of lye from the drain cleaning section of the grocery store. I had scored a gallon of methanol from a local chemistry lab. We followed Tickell's instructions carefully, although our scale did not measure accurately enough, and we made a batch of biodiesel.

It turns out that making biodiesel out of waste vegetable oil is not that tough. The process is called "transesterification," which entails taking one ester, vegetable oil, and transforming it into another ester, biodiesel. An ester is an alcohol molecule that is attached to a few carbon chains and looks sort of like a jellyfish. In vegetable oil the alcohol molecule is glycerin. When we make biodiesel we swap the glycerin for a different alcohol — usually methanol or ethanol — which means we put methanol in and get glycerin out.

The simple test for a successful biodiesel reaction is a nice clear line of separation between the biodiesel layer and the glycerin layer. The glycerin byproduct is a fascinating sidestream that is nontoxic and can be composted.

Despite our crude measuring technique and some glaring errors in our process, Gary and I made biodiesel on the first try. Because our fryer oil had been used only once, and because cooking a turkey requires that the oil stay at a relatively low temperature, we had bright yellow biodiesel with dark brown glycerin at the bottom of our mason jar.

Success was ours.

 *two*

Off to College

Whenever I am waiting to use the bathroom at the General Store Café in Pittsboro, I familiarize myself with the postings on the community bulletin board along the aisle. It's a diverse collection. Muralist seeking work. Tree service. Lawn service. Wild animal removal service. Crystal therapy, aromatherapy, massage therapy, talk therapy and colonics, which sometimes are offered at a discount.

One flyer that caught my eye was a black and white photocopy of a tractor in front of a silo with a headline that read: "Think Fuel from Vegetable Oil is Impossible?" It was advertising a new course on biofuels at the Pittsboro campus of Central Carolina Community College. I was intrigued and torn. Perhaps a class would help me get beyond the blender. Or perhaps it would slow me down.

Two adorable instructors, Rachel Burton and Leif Forer, immediately eradicated all my fears. They were in their twenties, they were enthusiastic, they were going to buy Joshua Tickell's *From the Fryer to the Fuel Tank* and they were going to make a batch of biodiesel in a blender.

I later came to learn that each of them had a long and diverse relationship with biofuels, but on that first night, having already read the book and having made a batch at Summer Shop, I couldn't help thinking that Gary and I were way ahead of them. What intrigued me was that these two were about to embark on the creation of a large-scale biodiesel processor and were willing to take the class with them. Lack of knowledge, experience or resources did not dampen their spirits a bit.

Leif had moved to North Carolina from the Northeast specifically to teach this class. He had been active in the founding of Greasecar, a small company selling kits that allow the user to burn straight vegetable oil in a diesel engine. He had obtained grants for his small private college and had designed and built a reactor, but had never had a chance to give it a spin. I think he yearned to travel back to Amherst, sneak into the basement where his virginal unit was stored, steal it and bring it to North Carolina for its maiden voyage. His work in biodiesel also had taken him to Africa and parts unknown. Leif was unassuming, quiet, relaxed and perfectly willing to intersperse his speech with words such as "like" and "um."

A memory still seared in my consciousness is Leif queuing up a video — *The Prize*, based on Daniel Yergin's book — and saying, "Um, this is a video about the oil crisis in the seventies and um, like, I wasn't really even born yet, but like, here it is …." I was both horrified and exhilarated when I realized that I was "The Old Man of Biodiesel." Rachel still uses that moniker when describing me from time to time, although it turns out to be wholly inaccurate.

Her journey into biofuels began during her activist past when she read an article in an occasional zine called *Live Wild or Die*. The article, on do-it-yourself vegetable fuel for diesel vehicles, hooked her imagination. She managed to land a free education in auto mechanics and wound up teaching automotive technology at Central Carolina. Mastering vehicles, it turns out, was simply one step toward mastering biofuels.

Nowadays when they want a night off they bring their biofuels class out to the refinery and I sub in as "guest lecturer." At one point we thought we could have the class make large batches of fuel but there are way too many people in the classes now. They barely fit into the refinery, and I am forced to hold the discussions on the cement pad in the backyard.

It turns out the College is our canvas. It is a resource that we have been able to shape. Not only does it have automotive bays with lifts that are handy for straight vegetable oil conversions but it also has lab space where students can perform titrations, make mini-batches of biodiesel and analyze the fuel.

And the College has a foundation that is set up to receive grants and charitable donations. As we were finding our way through the labyrinth, I drummed up a local donor who wrote a check for $7,500 to get us started. That triggered the creation of an advisory board and a financial vehicle that Rachel has used to run workshops and pay for courses, travel and supplies.

When we ran into trouble disposing of our glycerin, Farmer Doug from the College's sustainable agriculture program came out to the refinery to show us how to do a proper job of composting. And it was through the College that we ended up with Oneas, whom we call Dr. Feedstock. He is a PhD agriculturalist from Zimbabwe who operates his own nongovernmental organization focused on agriculture in the developing world. He currently is conducting research on mustard seed and sunflowers as possible oilseed crops and doing experiments feeding our glycerin to goats.

Oneas is the main reason we have become a farm. Rachel threw her flock of chickens into the mix and he bought some goats. Farmer Doug tilled the front field and made real compost happen. Rachel refers to our operation as Piedmont Biofuels Research Farm and I managed to get us a farm serial number from the USDA.

Piedmont Biofuels grew out of those early classes at the College. While we benefited from the resources of the College, our relationship with it imposed some limitations as well. For instance, it would not allow us to buy and store 55 gallons of methanol. For that we needed to have a business, so we used my sculpture studio, Moncure Chessworks. The College did not have a good space for us to do reactor design-build, so we used Chessworks for that as well.

Without any intention whatsoever, the biofuels program started creating a market for both homemade and store-bought biodiesel. Piedmont Biofuels was created to meet that demand, and it came into the world as a co-op, without a profit motive.

Chatham County is on the edge of Research Triangle Park, which was designed so that academia and business could function together. The ideas born in the classroom could find their way into industry and translate into jobs and prosperity for the state. In our own small, accidental way, that is exactly how Piedmont Biofuels was formed.

Rachel shepherds our relationship to Central Carolina Community College. She has learned to administer grant monies and to fish out purchase orders for equipment and she guides Leif and me in our interactions with academia.

Rachel and I now co-teach an energy class at the College that focuses on renewable energy of all types. It originally was called "Energy Class," then "Understanding Energy," after the subtitle of Cari Spring's book, and one semester it was titled "Advanced Biofuels," which was a misnomer.

In my early days as an instructor, I wrote up lessons and story problems for the class to solve. One of my favorites came out of my earliest blog entries:

Night One

Energy is the capacity for doing work. There are different ways of measuring the amount of work we want to get done.

A BTU, for instance, is one match. (At this point I burn a wooden match from end to end.) One BTU is the amount of energy required to raise a pound of water one degree Fahrenheit. There also is a calorie and there is horsepower and don't forget miles per gallon.

We can tie all these measurements together if we want to. That's up to you. Since I have deep math anxiety, I fear measurements and would prefer not to go down that road. On the other hand I do have a story problem I would like solved:

There once were two brothers who grew up in a highly competitive family. The older of the two went out into the world and made his fortune, and when he saw what humans were doing to the planet he decided to pursue renewable energy. He passed the hat among family and friends and built a 1.8-megawatt wind turbine that would create electricity. The younger brother went out into the world and made his fortune, and when he saw what humans were doing to the planet he started making his own fuel out of recycled vegetable oil. How many gallons of homemade fuel does the little brother have to produce to equal the energy output of his big brother's wind turbine?

We solved the problem on Night Two. It turns out that the 1.8-megawatt windmill will produce about 5 million kilowatt-hours this year. That should make it simple. We know there are 3,413 BTUs per kilowatt-hour. It turns out there are 120,000 BTUs per gallon of biodiesel — the highest of any alternative fuel.

Why would the brother with math anxiety ever volunteer to teach this course? With the help of Scott and Peter we managed to come up with this solution:

1 kilowatt-hour = 3,413 BTUs.
5 million kilowatt-hours = 17,065,000,000 BTUs from the windmill per year.

1 gallon of biodiesel = 120,000 BTUs.
17,065,000,000/120,000 = 142,000 gallons of fuel per year.

There are other factors, of course. The backyard biodiesel operation has required very little investment in either time or money compared to the wind turbine. The wind turbine has a lot of embodied energy and cost about $3.8 million. The biodiesel refinery was cobbled together out of scrap and cost less than $10,000. My guess is that for the same price as the windmill we could assemble a chemical plant that could produce a million gallons of biodiesel per year. This estimate is based on a research plant I toured in Iowa that had an output of 875,000 gallons.

Bear in mind that what is not calculated here is how many wasteful homes the precious electricity is sent to prior to being frittered away. Also not calculated are the brutally inefficient automobiles that suck up the precious biodiesel prior to frittering it away ....

Energy Class has matured since then and I have taken to following "Rachel's Rule," which states that the best way to improve on Energy Class is to bring in a guest instructor. To a very large extent that is what we now do. Last semester we did a field trip to the cogeneration plant in Chapel Hill, and distinguished visitors from NC State and the State Energy Office as well as private contractors have visited us from the worlds of geothermal and solar thermal.

Rachel's vision is to create a sustainability program at the College. It already has a vibrant sustainable agriculture program that brings students from far and wide. The College skirts around the edges of green-building courses and workshops and has some sustainable landscape design and some courses on wastewater treatment.

Rachel wants to broaden the feedstock research and has proposed a course entitled "Feedstocks and Sidestreams," which would examine alternatives to soy and explore options for the glycerin and the wash water that come out of the biodiesel process. She also is proposing a new class in biodiesel analytics under the mantle of bioprocessing, and she wants to brand our energy class as "social science."

The idea that this little college could assemble a two-year Associate in Arts degree in sustainability, or in biofuels alone, is not the least far-fetched. Rachel tends to get everything she wants, and if she wants it, it will happen.

 *three*

Breaking the Blender Barrier

Anyone can make biodiesel in a blender. The recipe calls for some dangerous ingredients such as methanol and lye. All are toxic, nasty substances in themselves and are even worse when combined in the biodiesel reaction. If you are interested in attempting to make your own bio-diesel, you should have chemical-resistant gloves, clothing protection, eye protection, fire protection and the ability to hold your breath for a prolonged period of time.

Apart from that, it's easy.

Get some methanol from your local race car shop or chemical supply shop or chemist. Get some lye from the drain cleaner section of your grocery store. Get some vegetable oil from the next aisle over.

Put 200 milliliters of methanol in the blender. Dump in 3.5 grams of lye. Blend. This is the first chemical reaction necessary for the process. The result is methoxide. The reaction needs to be complete. Stop the blender and look for unreacted granules of lye. Blend some more. A good methoxide reaction will heat the plastic or glass sides of the blender.

Methanol is a dangerous alcohol that comes from denatured wood chips or, more commonly, from natural gas. Although it can be produced from biobased processes, in the United States most of it comes from natural gas and it is a fossil fuel.

Unlike ethanol, which can make you "happy" and cause you to lose your inhibitions, methanol makes you go blind. Both chip away at your central nervous system, but methanol is much less fun. It's a dangerous substance that can permeate the skin. Its fumes are explosive and dangerous to breathe, and because it is soluble in water, it can threaten groundwater drinking supplies. Methoxide is even worse. It has all the qualities of methanol, but it burns the skin like furniture stripper as it passes through the membrane into the bloodstream.

Once you have a successful methoxide reaction, add a liter of vegetable oil and blend for about fifteen minutes. This is the biodiesel reaction, and if the mixing is done correctly you will get two nicely defined layers. One is glycerin, the by-product of the reaction, and one is biodiesel.

The glycerin is nontoxic and can be composted and the biodiesel can go right into the fuel tank. Many demand that the fuel be "washed" first to improve its quality, and washing is a wonderful debate in the world of homemade biodiesel.

In the biofuels class we all made blender batches. We didn't use just "virgin oil." We used grease from all over the area, including some straight sausage fat from behind Pepper's Pizza, a popular student pizza joint in Chapel Hill.

The process for waste vegetable oil is a little different from that for virgin, as the 3.5 grams of lye is no longer a constant. With each batch of fuel from waste oil, a titration needs to be performed to determine how much lye to use.

No one listened when I suggested that fifteen minutes was a little long as a duty cycle on a blender. I told Leif and Rachel that blenders typically run for seconds rather than minutes; after all, how long does it take to crush ice for a daiquiri? I like Martin Stenflo's quip that biodiesel is a great way to ruin a perfectly good daiquiri. And I would add that once the motors start smoking and the gaskets start giving out and the plastic starts cracking, the fact that the blender is a rotten tool for making biodiesel becomes apparent. Used blenders are ubiquitous at thrift stores, however, and consequently they are a "consumable" in the backyard biodiesel process.

The blender is at the heart of Joshua Tickell's successful popularization of biodiesel. The problem is scaling up the process to make more meaningful quantities of fuel.

Our first stab at this was done in class as a group, and it became an object lesson in how not to proceed. All of us were biodiesel enthusiasts. Some of us were renewable energy fanatics. Others had little practical experience. Some were mostly cerebral, with little real relationship to the physical world. But just the same, away we went. And in no time we had proposed pedal power and gravity feeds and preposterous ideas of how a biodiesel reactor could be assembled and would work.

I had a vision of portability and kept insisting that it be built on a trailer. I wanted a 100-gallon chemical plant that people could borrow to make their own fuel. My early thinking was along the lines of the portable sawmill I had once brought in after Hurricane Fran devastated our farm. The budding activist within me wanted to be able to make biodiesel on the courthouse lawn.

We collected vessels and bicycle parts and containers and motors and pumps. If the blender could be called "Reactor 1.0," our attempts at 1.1 were a disastrous example of groupthink. We had assembled a car carrier, a large double-axle trailer that seemed to have enough space for all our designs. On it sat a 163-gallon elevated aluminum tank with a PVC ball valve mounted near the bottom. We intended this tank to be our reactor vessel.

One of the tank's myriad flaws was our inability to know how much veggie was inside it. Leif set about solving this problem with a simple sight tube. He picked up a clear plastic piece that had no relationship to PVC and we spent a week attempting to plumb it to the unit. Once it was in place, I went to calibrate it. My nephew David was in town at the time and was eager to help. I'm not sure how much he weighed, but I would guess around 120 pounds.

We carefully added a gallon and marked the level on the sight tube. Then another gallon, slowly, exactly, drawing indelible lines on plastic. Delighted with our progress, I stepped back to admire our work. David jumped off the trailer and every calibration went out of whack. The trailer was not level when we started — one of its wheels was rather in a ditch — which meant that our accuracy was dependent on getting the trailer into the same ditch and locating a 120-pound nephew in exactly the same spot every time. Rachel relishes the memory of the day we threw out the first sight tube.

When the first semester ended we essentially had nothing to show for our efforts but an assortment of flops and bad ideas. The Christmas break put us out of our misery, and the only people still interested in design were Rachel, Leif and myself. That simplified matters greatly, although I suspect the best size when it comes to backyard biodiesel design probably is a group of one.

We continued doing our design work at Chessworks, but after a few vegetable oil spills I realized I was spending an inordinate amount of time cleaning the shop. Spilled vegetable oil was incompatible with sculpture sales, and we evacuated to Summer Shop.

Summer Shop is fifty yards from my house and is where my sculpture business began. It is merely a metal pole building with a gravel floor and some walls made out of scrap tin. It has very little space that stays dry in the rain and only occasional heat from a burn barrel of waste construction debris. At the time, it had no running water.

Primitive as it is, it is awash with creative vibrations. Vestiges of the sculpture business adorn the woods, and early pieces, flops and an amazing scrapyard surround the place. It is an incubator, a laboratory and a wonderful place to work.

At Summer Shop we went to work on the 7,200-rpm motor that Rachel had scrounged up. I took a propeller I found at the scrapyard, welded it to a shaft and attached the shaft to the motor. The shaft and propeller weighed about ten pounds, but the motor rotated it easily. We probably could have water-skied behind such a device.

We essentially were building an upside-down blender: steel drum, motor mounted on top, shaft and propeller protruding downward into the vegetable oil.

We were mindful of safety. One of the policies we implemented, for instance, was a "drink station," where water glasses and coffee cups were parked. We had decided that if all drink containers were limited to a single place, there would be less chance of someone gulping down lye water, isopropyl alcohol, methanol or the like.

One chilly fall morning, Leif and I filled our untested unit with waste veggie, added the methoxide and fired up the reactor. It sounded like a jet airplane taking off. We were stoked. We went around the corner to the drink station for coffees. As we conversed, the propeller shaft bent under the force of 7,200 revolutions per minute submerged in 400 pounds of liquids, and when the stainless propeller brought the steel drum up to "cherry red," a good-size explosion occurred. It shook the windows at the house. Janice heard it through the woods, a half-mile away.

We easily put out the ensuing fire and headed back to the drawing board for what would become reactor 1.2. Some Internet research led us to a chemical mixer that could mix up to 100 gallons at a time. It was 350 rpm and had a couple of propellers the size of silver dollars. I was amazed at how delicate it appeared. We hooked it up atop a plastic 55-gallon drum from the scrapyard and our upside-down blender was perfected.

By this time our "burn rate" for grease was accelerating. We would travel to a restaurant, scoop out 30 or 40 gallons, bring it back to Summer Shop, attempt to make a batch, get the recipe wrong and make 40 gallons of soap. That would trigger yet another grease-run, which we would bring home, react and throw away. Two years later Summer Shop still has vestiges of bad batches and flops around the place.

Leif took off for the Christmas break and Rachel and I kept working. When the weather turned cold, we put on more clothes. Because of the methanol, we were afraid to run the burn barrel. We would start each cold morning early, firing up propane heaters and torches to transform the used fryer oil into a liquid state.

We made mini-batches. We fetched more grease. We spilled. We labored. We lifted. We cleaned. We got to know one another. And we longed to see separation between the fuel layer and the glycerin layer in a big batch. Eventually some basic wisdom kicked in. Rather than doing experimental runs at full capacity, we backed our batch size down. Thirty gallons became twenty, which became ten. And we had no luck at all. We could make endless successful mini-batches in a blender, but nothing more than a quart of fuel.

Christmas Eve was upon us. Rachel had canceled her travel plans. I ignored my children. We did not work on Christmas Day, but we were back at work the day after. We were obsessed with making fuel.

Our breakthrough came via heat from the wood-stove. Tami and I were headed out for an evening on the town. Rachel and I made a batch, scooped five gallons out of the bottom of the reactor into a clear hard-plastic container that Leif had left behind, carried it up to the house and slid it into the corner of a kitchen counter. The kitchen was warm compared to Summer Shop. Rachel took off, I went on date night and when I came home two distinct layers had formed.

News that we had broken the blender barrier spread up and down the eastern seaboard. In no time, a group of biodiesel enthusiasts from Pfeiffer University (over three hours away) were gathered around Summer Shop. We would make ten gallons of fuel and there would be six people standing around celebrating. Everyone would get less than a gallon and a half to get home on.

We had a group from DC come through with a hand-held video camera to get our every move. It turns out we were not alone. Everyone was trapped in a blender. Everyone wanted to make meaningful quantities of fuel.

We now know that most backyarders shut down for winter and that fuel making is largely a three-season activity in our climes. I talked to one backyard brewer at the Shakori Grassroots Festival in mid-October. He had battened down for the winter, putting up 90 gallons to see him through until spring. Piedmont Biofuels now is one of the few small producers in the region with indoor capabilities allowing us to make fuel year-round.

Once we had ten gallons perfected, we started stepping our quantities up. Reactor 1.2 was reliably pushing out 30 gallons of fuel per batch before it was moth-balled for a superior design. I regret that Leif was not around for those early batches, because the "start small and step it up" wisdom that Rachel and I had gained the hard way was lost on him, and the lesson came around again a year later with methoxide storage.

There are two schools of thought on methoxide. Some, like Joshua Tickell, claim that it does not store well and needs to be made and used right away. Others claim a nice long shelf life. Call me Tickellesque, but I subscribe to the "make it and use it" school.

On one occasion Leif decided to pre-make methoxide and store it for a couple of days before doing the biodiesel reaction. By then we had moved to reactor 1.5 (the one that came to be known as "Alien Baby") and Leif was routinely turning out 70- and 80-gallon batches. I tried to convert him to the "step it up" philosophy, but he ignored me and made 70 gallons of soap with stale methoxide.

Not everything we learned during those early days of trial and error made it into the canon of biodiesel wisdom, but we have been showing people how to break the blender barrier ever since that first ten-gallon batch.

And I have been publishing some of our design breakthroughs in Energy Blog along the way:

First Ride of the Salad Spinner

When the biofuels program down at the College was in its infancy, the whole class was involved in the "design" of our first reactor. People were enthusiastic. Ideas swirled about in abundance. And the junk flowed in from several directions. There was pedal power. Passive solar. Active solar. 12-volt from batteries charged by the sun.

In the middle of it all, Rachel and I took a left turn. We were tired of flawed concepts, weary of bad ideas and beaten down by notions that came from inventors who had never actually produced any inventions. Our departure from "groupthink" occurred one brisk fall afternoon at Chessworks.

We took a plastic 55-gallon drum with a lid, an agitator that we liberated from a washing machine and a circular handle. We added a piece of aluminum pipe and in two elegant pieces created the "Salad Spinner." Powered by the human hand on top, a simple cranking motion caused the agitator to spin around and around and move a huge volume of water. We cranked backwards and were in awe of the easy "mixology."

The group rejected our work, saying it would never hold up to the rigors of biodiesel production. We reluctantly agreed and set it aside. The 55-gallon drum later became the basis of our first 30-gallon reactor, and the remains of the Salad Spinner sat, untried and unproven on the back wall of Summer Shop. That wall is an interesting place. I like to think of it as an island of misfit designs. Others have referred to it as the "Great Wall of Flops."

A couple of years later I was smitten by a methoxide mixer created by Tom Sineath over at T.S. Designs in Burlington. He's an inventor whose creations see the light of day. They are not just talk over at T.S. Designs. I've never asked Tom whether he has a "Great Wall of Flops," but my guess is that my collection is not unique. They sent an e-mail picture and I was startled to see a mixer that was basically our Salad Spinner. I congratulated them on their genius and regretted the early days when our group of designers was significantly larger than it is today.

Yesterday Rachel, Scott and I made an 80-gallon batch. Rachel was in the middle of titrating and was complaining about our dangerous and "janky" methoxide mixing methodology. After all, we still are hovering over 15 gallons of methanol and lye with an electric drill and a drywall mud mixer. We can see sparks.

In a moment of inspiration, Scott and I drove to Summer Shop, retrieved the forgotten elements of the Salad Spinner, carried them back to the refinery and scrubbed them spotless. By the time Rachel had the recipe set we were ready to go. In went the methanol and the lye, and we proceeded to spin and spin.

It was not fumeless. From a distance you could see wisps of fumes from time to time, which I suppose precludes my idea to attach a Playskool Sit'n Spin atop the unit and get some little kid to climb on and do our mixing for us. But it was effective. We got a great methoxide reaction with very little effort, we removed an electrical step, thereby helping our energy balance, and we made our whole process safer.

We have attempted to stay focused on energy balance and I believe the class at the College has helped keep us mindful of BTUs. Leaving the blender behind is at the heart of every backyarder's journey, and reactor designs vary widely. Our obsession with energy balance has left many designs out of our thinking. For instance, electrically powered aquarium heaters are commonly used, but the idea of using several hundred watts of electricity for prolonged periods to bring oil up to temperature is anathema to us. We are forever trying to whittle away at the fossil components of our fuel, and our understanding of energy balance is an important part of our story.

At one point I published an analysis of our homemade fuel in the blog:

Energy Balance

I have been curious about the energy balance of our biodiesel operation. For each unit of energy we put in, how many are we getting back? I once posed this question to the Biofuels list and the answer that came back was to "try Homer."

I did that today. It is an amazing tool. It's a Windows application published by the National Renewable Energy Lab (NREL) that allows you to model energy usage across a huge array of platforms. It is packed with information. It automatically graphs efficiency and allows you to modify inputs. Very cool.

I wanted to explore the impact of moving our refinery to a biodiesel-powered generator. My starting point is that biodiesel made from virgin soy has an energy balance of 3:1. For every BTU of energy that goes into its production, three come out.

That ratio includes the growing, harvesting and pressing of the soybeans, with all the petrochemicals, pesticides and such that are used along the way. After the precious oil has been put onto petroleum-powered trucks, brought to the restaurants of North Carolina and used to cook fries, we come in.

We travel about 45 miles to get our used cooking oil. Let's say my 1992 Dodge pickup gets about 15 miles to the gallon. That's three gallons of B100 each way or six gallons round trip. What we pick up varies. Let's say we pick up 50 gallons of veggie each time and make one run a week.

When we make biodiesel we typically use between 15 and 20 percent methanol to get a good reaction. Let's say 20 percent. And we use electricity in our process, probably about 250 kilowatt-hours per month.

I am writing this, by the way, at Glen's frigid house on Georgian Bay. I am performing these calculations at his kitchen table and he is under the misguided notion that spring has arrived and it is time to grill some fish.

I'm going to skip the propane we use, since it has virtually been replaced by solar thermal and the weather is changing in North Carolina. Tami tells me the boys were out picking flowers today. Hyacinths and daffodils and crocuses. From what I can tell tonight, Glen will not see a flower for months. He's on the porch right now, firing up his propane grill to prepare a piece of fish that came out of the bay across the street.

I wrap another blanket over my shoulders, curse his "wet" firewood supply and thank my lucky stars that I am not trying to make biodiesel in this forsaken place. "How many BTUs per kilowatt-hour?" I ask.

He's excited about grilling outdoors tonight. "3,413," he says. "I couldn't have found the grill under the snow last week."

I'm trying to find the energy density of methanol on the net. Glen comes in with a bottle of olive oil he has had on the porch. Holds it up to the light and says, "Check it out — no problem with flow." The bottle is starting to accumulate frost. We have spent the afternoon discussing the inferior cold flow properties of waste vegetable oil.

Back to energy balance. We require 24 gallons of biodiesel equivalents per month to procure our feedstock. We use 6.65 gallons of biodiesel equivalents in electricity and another 17.7 gallons of biodiesel equivalents in our methanol. That's 48.35 gallons per month needed to make 200 gallons of biodiesel. Divide one by the other and you get an energy balance ratio of 4.14:1. For every BTU we put into making fuel we get 4.14 BTUs out.

That's a disappointing model. If virgin product is 3:1, we are not markedly better. But I have been free and easy with the assumptions. If we sell a vehicle at Car Max when we pick up the veggie, where do we charge the energy? Sometimes we do the grease-run when we are in town anyway. If we used Tami's Jetta to pick up grease, our energy balance would be much improved. But if we did that I would be single, and how would we calculate the energy balance in that?

~~~

Nowadays when people ask me about reactor design I send them straight to girl Mark's fumeless processor, which she assembles out of old hot water heaters. It is the ideal vessel since it can easily be pressurized for methanol recovery, it is safe and it has a ready way for heat to be added to the biodiesel reaction. Girl Mark has a wonderful book on backyard for sale at <www.localb100.com/book.html>.

Also available is the reactor design section of <www.journeytoforever.org>, which offers a multitude of safe designs that can help backyarders break the blender barrier without all the dangerous grief we encountered.

four

Do-It-Yourself Ghetto

I once wrote an article for *Private Power Magazine,* sort of the Canadian equivalent of *Home Power,* and the entire time I was writing it there was an audience of "do-it-yourselfers" in my head. It wasn't as bad as the time I wrote an essay on outhouses for *Harrowsmith,* when my brain was full of "back-to-the-landers," but it was close.

As I wrote for *Private Power* I started questioning the massive amount of do-it-yourself required in backyard biodiesel. If you don't like to plumb things, pass on building your own biodiesel setup. Welding also is helpful. Carpentry too. Successful backyarders tend to be handy and to have a lot of tools. The movement is populated by expert scroungers, recyclers, inventors and generalists who put the average homeowner to shame. A great deal of our homegrown invention is chronicled in the blog, as in this entry, in which our struggle is evident:

Pants First, Then Shoes

I once saw a cartoon — I think it was a Gary Larson "Far Side" — that had a fellow sitting on the edge of his bed, staring blearily at a sign that was sticky-taped to the wall. The sign read: "Pants first, then shoes." I'm not sure why that cartoon stuck with me, but I have remembered it many times.

There was the day Leif discovered the chain hoist affixed to a girder at Summer Shop. The hoist could easily lift a full 55-gallon drum of methanol, around 400 pounds. He was so enthusiastic about his encounter with a new tool that he decided to put a couple of drums on the back of his Mazda and run them over to the new refinery.

We have to thank Leif for that, except when he arrived at the refinery there was no way to get the drums off the truck. No dock, no elevated structure to attach a chain hoist to, nothing. We tried hanging chains off the bough of a nearby hickory tree and swung some full barrels of methanol around in the breeze, but nothing seemed to work. It was a dangerous, stupid folly that could have been averted with a simple sticky-taped sign: Pants first, then shoes. Don't put the methanol on the truck until you have a way to get it off at the other end.

On this occasion I resorted to building a makeshift dock out of old tires and rolling the drum onto the bouncing pile. This method is recommended only for those who are interested in poisoning the groundwater, incurring serious nerve damage or renewing their intimacy with their chiropractor. It was kind of like trying to pull a tight pair of pants over steel-toed boots.

We are on the Eve of California tonight. Tomorrow we begin the "Shadow Conference" on biofuels. A couple of weeks ago, as Rachel and I were approaching the counter at the Jordan Dam Mini Mart, she was railing at me about washing our fuel. I could see the apprehension in the clerk and the nearby customers as she barked, "We don't wash. I don't want to show up in California unwashed. I don't want to face them and say we don't wash ...."

Rachel had it right. The next day we agreed to draw five gallons of fuel out of Alien Baby and run it through the wash cycle. We had the aquarium pump and bubble stone, and we used a plastic container with a spigot on the side. We added a few gallons of water and watched it bubble away. After a night of bubbling, there was a little bit of scum on the top in places, but we hadn't created a batch of soap.

We were pleased until we realized that we had an 80-pound container that would need to be propped in such a way as to keep the rubber seal at the spigot dry. We had created a large vessel of washed fuel that took a ton of space and would spill on a moment's notice, and we had not thought of how we would separate the wash water from the fuel. It was like trying to pull sweatpants over hiking boots.

We have not yet perfected our wash station. My guess is it will be a 110-gallon conical vessel with a stand of some sort and preferably with a garden hose to the compost. Surely someone in California will have ideas on that.

Last night I drained the washed fuel into two five-gallon plastic grease containers I had managed to clean. They are made of an opaque high-density polyethylene (HDPE #2) plastic and the wash layer became readily apparent. While it is always cool to see separation, it was a great relief to have the fuel safely set aside, labeled, with a lid on it.

I thought about the cartoon. Don't wash your fuel until you have a way to drain the wash water out of it. At least when we show up in California we can say we have washed ....

I continued the washing thread much later, but it still smacks of a heavy dependency on the ability to "do-it-yourself:"

Double Bubble

We are washing fuel tonight. And when I say that, I don't mean some little aquarium pump pushing air through a spent grinding stone. This is the real thing. Today Leif rigged up Wash 1 and Wash 2 with some serious bubble action from gear he bought from Aquatic Ecosystems.

When I left the refinery tonight, he had 80 gallons roaring away in the most elegant bubbling arrangement I could imagine. Rachel named it "Double Bubble" and it is an apt name. Double Bubble consists of two giant yellow drums that Leif bought at a surplus sale for six bucks. We are unsure of their size.

When we were in California at the Shadow Conference we picked up a decidedly "anti-plastic" bias so we plumbed our huge drums with metal fittings. They appear to be stainless ball valves in brass housings and they are beautiful to behold.

When I arrived at the refinery tonight, Leif and a buddy had a big batch bubbling in Wash 1, which is elevated on cinder blocks. They had constructed a shelf, bolted a pump to it and dropped clear plastic air hoses into foot-long bubble stones at the bottom of each tank.

Apparently the pressure was too high, and the initial offering brought the fuel to a robust bubble. Each vessel can hold 80 gallons of fuel and 20 or 30 gallons of water. The water is heavier than the fuel and falls to the bottom, where it picks up an air bubble that carries it back to the top of the batch. Then the bubble breaks and the water falls to the bottom again. Along the way, the fuel is gently scrubbed. Unreacted methanol, unreacted lye, ash, glycerin and soaps are scrubbed out during the wash step.

Each vat is covered by a 33-inch plywood lid, which Matt and I made up at Chessworks the other day. We grabbed a chunk of ¾-inch plywood (that once formed the backdrop for sculpture) from the scrapyard at Chessworks and rendered a couple of circles out of it. I found a pair of stainless rings, which I operated on with the plasma torch to create the most magnificent lid handles imaginable.

It's funny. It appears we are leaving the "unwashed masses." Now that Double Bubble is set up, we probably are scheduled to become "washed fuel" snobs ....

As long as we all have scrapyards and the ability to manipulate plastic, wood and stainless steel, we are set to go into biodiesel production.

Part of me applauds the work of Rudi Wiedemann, who assembled his Fuelmeister and got it into Real Goods for $3,000 a pop. It's way overpriced and I doubt he will ever have a repeat customer, but that is beside the point. What I like about the Fuelmeister is that it exists. Reactor designers from around the world love to savage it online for its shortcomings and I can find fault with it too, but Rudi had the courage to put it out into the world as a product, which means that those who fear designing and building reactors themselves can click on a mouse and have a Fuelmeister show up on their doorstep.

I think about my older brother, Jim, a competent handyman who has done more than his share of building, refinishing, restoration and remodeling. He is an engineer with a thorough understanding of how things work but his schedule prevents him from spending endless winter nights tweaking propeller blades and producing flops until he gets it right.

Jim is interested in his own ecological footprint and in sustainability in general. He might have time to assemble a kitted biodiesel plant that he can buy online but he doesn't have time to do everything himself, so he buys a hybrid vehicle and calls it a day.

And the same logic applies to active solar and solar thermal and micro wind and the entire renewable energy sector that is so easily encapsulated in magazines like *Home Power* and *Private Power.* Don't get me wrong. I can settle in at the kitchen table and argue fittings and sight tubes with the best of them, but I wonder if in the process we are leaving the majority of people out of the sustainability equation because we demand they do everything themselves. Perhaps renewable energy in all its forms is too closely tied to an individual's ability to do the work and that connection could leave sustainability in a DIY ghetto.

I posed this question to my Energy Class down at the College, and the overwhelming consensus was that having people do things themselves gives excellent feedback, that it engages them in the process of living in a sustainable way and that there is an inherent flaw in trying to "make it easy" for people like my brother Jim.

I'm torn on this issue. I like to do things myself and I believe in feedback loops for everyone. When we have feedback on the effort required to make a gallon of fuel or to bring the kitchen up to temperature with the woodstove, we have a better chance of modifying our behavior in the name of sustainability. At the same time, if too much effort is required a lot of people are left out. Jim might not mind heating with wood but should he have to design and build his own wood-stove?

I don't like to limit this conversation to biofuels. If you want to go solar you'd best familiarize yourself with watts and amps and voltage. If you want to do pas-sive solar you'd best learn about zones and angles and degrees. If you want an electric car you really need to convert an existing vehicle to electric power. If you want to run on straight vegetable oil it's best to be a shade tree mechanic.

The relative newness of the biodiesel movement means there are tremendous gaps in the knowledge base. Leif and Rachel, for instance, are dying for a compre-hensive textbook on biodiesel from which they can teach their classes, but if they really want such a thing at this point, they will have to write it themselves. In the meantime they live on large PDF documents like the ones Jon Van Gerpen has published through the National Renewable Energy Lab, or on zines like the

Biodiesel Homebrew Guide, which girl Mark publishes whenever she gets near a photocopier.

Gerhard Knothe, Jon Van Gerpen and Jürgen Krahl came out with *The Biodiesel Handbook* in the winter of 2004, but at $100 a copy it is as expensive as a 16-week course at the College and didn't find its way into the B100 Community in quantity.

Because of the information shortage associated with biodiesel — which includes feedstock and sidestream research we are forced to conduct ourselves — a certain amount of do-it-yourself is still required. That's both exciting and frustrating, and I sometimes find myself wondering when we will break free of the DIY ghetto.

 *five*

Straight Vegetable Oil

Whenever the news crews want to cover alternative fuels, they gravitate toward those who have converted their vehicles to run on straight vegetable oil. It's wacky enough to get noticed, and the public often is introduced to biofuels by a bunch of nutty kids in a bus powered by vegetable oil.

Converting a vehicle to straight vegetable oil (SVO) is simple. Merely get an additional fuel tank with a radiator pre-installed, cut into the coolant lines under the hood and redirect them to the radiator in the extra tank. That way the waste heat generated by driving down the road is moved into the vegetable oil to make it good and hot. By simply "teeing" into the fuel system and adding a solenoid and a switch on the dash, the driver is able to tell the system to "stop sucking from my diesel or biodiesel tank and start sucking from my vegetable oil tank."

When I was running on straight vegetable oil I published a blog entry revealing the point of view of my son Arlo when he was six years old:

Rocket Fuel

I drive a 1992 Dodge pickup that was converted to run on straight vegetable oil by the biofuels class at Central Carolina Community College in the winter of 2003.

It's a loud, old, inefficient pig that once was used for hauling greyhounds from race to race and was modified by the original owner. There are a bunch of switches embedded in the dashboard. One tells my fuel pump where to suck. When I throw that switch, a light comes on and soon my exhaust starts smelling like french fries.

Each day I take Zafer and Arlo to school. We start on biodiesel and roar down the Pittsboro-Moncure Road. When I think the vegetable oil is hot enough, I turn to the boys and say, "Should I switch to veggie now?"

They generally protest loudly, often chanting in unison, "Diesel, diesel, diesel." And I think it's because they don't want us smelling different from the other cars that are pulling through the parking lot in front of the Moncure school. I like smelling different and always throw the switch so we are pulling in on pure veggie.

Last night I took the family to the final biofuels class of the fall semester over at the College. It was a potluck affair, very relaxed and casual, complete with a student presentation on a continuous-flow reactor design to be implemented in the backyard.

The boys ran wild as usual, and when class ended we headed to the General Store Café. Surely the highlight of the evening was when Rachel got up in front of her new crop of biofuels graduates and awarded Tami her honorary diploma. It read: "33 Contact Hours" for "Biofuels, Biodiesel and SVO."

It should have read "3,000 Hours." Tami has watched the biofuels project grow up in our backyard. She was taken aback when I used her washing machine to clean a load of greasy sock filters. She has fed biodieselists who have wandered up to the house from Summer Shop. She has picked up after them. She has made them thermoses of coffee on cold winter mornings and shared leftovers with passersby. She has fetched towels for weary biodieselists who have jumped in the hot tub and she has laundered the clothes they have left behind.

On the edge of a mountain in Nicaragua last spring, she looked down through the morning mist on the canopy and listened to many of us rave about sustainability, and when she came home she shed her SUV and bought a diesel Volkswagen Jetta wagon, which she has routinely filled up with B100 throughout the year.

If we could make more, she would use more. In many ways that decision of hers inspired the group as a whole.

I know it inspired me. Tami is a motherly figure in our bio-fuels project. She doesn't clean used vegetable oil from the bottom of 55-gallon drums and she doesn't scoop out of Dumpsters, but she has an enormous presence.

Last night as the festivities at the General Store were winding down, I loaded the boys into the Dodge to head for home. Tami came out of the store behind us, and suddenly the race was on. "Step on it, Dad. We can beat her!" they screamed.

As we started down the road I heard Arlo say, "Which is faster, diesel or biodiesel?" He stretched out his seatbelt and positioned himself next to the various switches on the dash.

"They are the same," I said.

"Which is faster, biodiesel or straight vegetable oil?" he asked.

"They are the same," I said.

He was clearly distressed by my answers. "What's faster than all of them?" he said.

"Rocket fuel" was my answer.

At that his eyes lit up and he started throwing random switches on the dash. "Switching to rocket fuel!" he proclaimed, as we lumbered into the night in pursuit of his mom.

Some people like to add an inline heater to the SVO system. Ed Beggs, up in Canada at Neoterics, sells an excellent 12-volt inline heater he calls a "vegtherm."

One of our group, Peter Denz, designed his own using a glow plug to heat the veggie prior to combustion.

One catch to the two-tank system is that you must start your journey on biodiesel, get the motor good and hot, switch to veggie and remember to end your journey on biodiesel. If you forget to switch back toward the end of the journey, to flush the fuel lines of veggie and fill them with biodiesel, you can have a very difficult time starting on the next cold morning when the veggie in your fuel lines has turned into a solid. This is not an issue in warm weather, when the engine will readily start without purging the lines at all.

I bought a two-tank system from Charlie over in Missouri, who runs a company called Greasel. Greasel then had all the attributes of a small, caring company. "You want to order one? Let me go out to the truck and find my order book" was Charlie's response to me over the phone.

And when the kit arrived (without a packing list), it was a simple cardboard box with a bunch of hose clamps, an expensive Racor reusable fuel filter, a big plastic tank with a radiator in it and Charlie's famous invention at the time, a "three-in-one" fuel line. Basically he wrapped the outgoing fuel line in some plastic to surround it with coolant lines so that the heat would travel with the fuel all the way to the injector pump.

The design was both interesting and lame. The three-in-one tends to be exceedingly hard to thread through the engine and firewall and usually ends up flayed, thereby defeating its purpose. However, Greasel is innovative, and Charlie needs to be applauded for his work. He returns phone calls, he pushes forward with research and development, and he supports his customers. Greasel is the type of small business that will thrive on the frontier of alternative fuels.

My two-tank system was by and large a disappointment. It was installed as a class project down at the College, and it was our first install. The fancy Racor filter started pulling a vacuum and halting the flow of fuel and there were constant air leaks in the system.

Diesel engines must be airtight to function properly. If air gets in the fuel system, fuel pressure is compromised, and without fuel pressure the engine does not run. There is no better way to introduce air into a stock diesel system than to cut into it and add an after-market vegetable oil system. Shade tree mechanics, who like to constantly muck around under the hood, can do well with two-tank veggie systems and can get a lot of miles on straight vegetable oil. People who do not want to mess with their vehicles constantly, beware.

I know a lot of people who have installed two-tank systems made by Greasel or Greasecar or homemade (most who buy kits end up modifying them). Some of them sing the praises of SVO and boast of many a care-free mile. Others confess that running on SVO can be a lot of trouble, downtime and work. Often when we hear from people on SVO, it's with problems they would like us to solve.

Science and industry are not yet ready for straight vegetable oil. They claim that regular ignition of veggie will coke injectors and lacquer combustion chambers, causing irregular spray patterns, loss of engine performance and reduced engine life.

The problem with their worldview is that most people who convert to straight vegetable oil are doing so on vehicles that have long passed their prime. When I started running my 1992 Dodge on straight vegetable oil, the possibility of long-term lacquering was the least of my worries. Before I delved into the joys of SVO I had 305,000 miles on the engine. I say that with sarcasm now, after Rachel ripped the entire SVO system out at my request. When she did so I wrote about it in the blog.

SVOectomy

The Dodge stopped starting. Drat. I took it to a non-diesel mechanic for an oil change and when they called me at the end of the day, instead of saying "You can come and pick it up," they asked, "What's the secret to starting this thing?"

I had it towed to Chatham Alignment. Those guys are the kings of all things diesel. Generally they are working on huge rigs. They have kept an eye on the biodiesel movement. They inspected the Dodge before I bought it and they restored our tank truck to road-worthiness after we acquired it. They passed on my Mercedes restoration project. They don't like $3.50 a gallon and they don't have the time to make their own, but my hunch is they get a kick out of the whole biodiesel thing, especially when they are called to tow one of our vehicles off the road.

So when my dead Dodge arrived, they took it in stride. They dug into the problem and ascertained that it was fuel-pressure related. They installed a new lift pump and mucked around a bit and got it to start again. When I arrived to pick it up, and to pay their handsome fee, they railed at my system and blamed it for my difficulties.

What they lack in social decorum they make up in technical expertise. This time they complained mightily and they sent me down the road with the single message: either redo SVO from end to end or rip it out entirely.

When the Dodge died again, they were on vacation. I had it towed down to the College so that Rachel could work her magic. SVO is for people who like to tinker, people with time on their hands. It's not good for mission-critical applications.

And my Dodge is critical. It hauls kids and veggie and methanol in equal parts, and when it is down for a couple of weeks the movement takes a hit. I need my truck. Downtime is no fun, which is why Rachel spent an entire day ripping out my SVO system and replacing all my fuel line with Viton hose as she knows my Dodge runs exclusively on B100.

I know a handful of people who have gone to the trouble and expense to convert to two-tank SVO, and almost all of them have abandoned ship. Most are running on straight biodiesel; some are so discouraged they run on petroleum. I have joined their ranks. Rachel gave the Dodge an SVOectomy. Now it will run only on B100. Now the fuel pressure is good and the truck starts every time.

One of the main problems with two-tank SVO systems is their application. They are a rotten choice for "around town" driving. Each morning I take the children to school. That's six miles, just about enough time to get the vegetable oil hot and flowing. But as I head to town, which is a ten-mile trip, I get only about five miles before I need to switch back to biodiesel. That's plenty to remember on one short trip, and if I fail to throw the switch I either have to drive longer or idle forever to clear the lines. If I were to start the truck in Moncure and drive to California without stopping, straight vegetable oil would be a good choice.

Apart from the two-tank systems, there also are single-tank systems in which the user can run on straight vegetable oil, diesel or biodiesel, all out of the same tank. The legendary version of this comes from a German company called Elsbett, although Neoterics also markets a single-tank solution.

Alexander from Elsbett conducted a couple of conversion workshops with Rachel and others down at the College. These workshops always receive rave reviews, and those who install the Elsbett system seldom complain of difficulties. That may be because the system is so expensive (they sell their wares in Euros, which drives me nuts) or because a German Elsbett engineer who has vast experience with all diesel engines installs them. Or it may be that the users I know are reluctant to report problems to me for fear I will publish the conversation in my blog.

Whatever the case, it is true that little can compete with the thrill of bombing down the road on straight vegetable oil. When the system is working, the driver is free. Free of an overarching fuel infrastructure that requires monopolies to deliver fuel. Free of Big Oil. Free of war. Running on straight vegetable oil does feel fantastic.

Used Veggie Collection

The starting point for collecting used vegetable oil usually is the front door of an eating establishment. Backyarders consider it bad form to poke around in Dumpsters without permission but permission is almost always granted. Many restaurant owners are surprised that someone wants their used vegetable oil.

Most people refer to used fryer oil as "grease," which is one of the many misnomers in the biodiesel vocabulary. People who suck grease traps for a living and dispose of "brown grease" consider used fryer oil to be a significantly higher form of life. And it is true that there is a mature grease industry outside of biodiesel. While biodiesel enthusiasts everywhere are quick to exclaim, "We are recycling a waste product," used fryer oil generally gets recycled anyway.

It is not really news that there is good energy left over in used fryer oil. The oil typically finds its way into animal feeds and makeup, and the route by which it does so is well-established. One of the players in the field is Griffin Industries, headquartered in Kentucky. They collect used fryer oil in Moncure, where our operation is, truck it to Charlotte, NC, sort it, blend it and dewater it. Then they ship some to Florida, where it is loaded on container ships and sent to Europe to be made into biodiesel. They ship some to Kentucky, where they make their own biodiesel. Some of their finished product ships back to North Carolina, where our Department of Transportation buys it to reduce the emissions from its fleet.

Griffin ships brown grease and yellow grease and white grease and fancy grease to customers all over the world. They joke that the stench of rancid oil smells like money to them. But they don't all have a good sense of humor about it. I once had lunch with their head chemist at the Iowa State Energy Center in Nevada, Iowa, where we both were studying biodiesel.

When he introduced himself, I lit up with recognition of his company name. "I've been in some of your Dumpsters," I said. His curt reply was "We press charges."

Stickers and warning labels on Dumpsters often indicate that the contents are the property of the hauler and not the restaurateur, so the permission easily granted at the front door of the establishment may be meaningless. I wonder if the Dumpster haulers are aware of how often their warnings are ignored.

The used fryer oil market varies from one locale to another. In North Carolina some people pay to have their used veggie hauled away. Some have it toted away for free and some get paid for it. As a result, some value it and others don't.

Rachel and I encountered a prime example of this on a grease-run in Sanford, North Carolina. Our first stop was at Hardees, a fast-food chain known for its burgers. The manager told us to help ourselves to all the grease we could carry. We opened the unlocked fence to behold a splattered and open grease Dumpster with chunks dripping down the sides. We saw a rotary pump that appeared to be used to move grease and some buckets of water and slop. The place smelled horrific. We hand-rolled 30 or 40 gallons into a barrel on the back of our truck and decided we'd had enough.

Before departing we walked across the parking lot to a Bojangles, a chain known for its chicken. One of their slogans is "Carolina Born and Breaded." We decided to check out the grease before talking to the manager and found a locked pressure-treated fence around the Dumpster. I climbed over. Inside was a spotless Dumpster with a watertight lid. I lifted the lid and saw the most beautiful sea of grease I had ever beheld. It was orange and clear, dry and perfect. When we went in to ask the manager for permission to carry away some of his used oil, he told us to forget it. Clearly one fast-food place had discovered the value of its sidestream while the one next door had not.

Having explored the interiors of many Dumpsters and barrels in Research Triangle Park, I can say with authority that there is not always a connection between the quality of a restaurant's food and the quality of the vegetable oil out back. For some reason, the Chinese food place in the strip mall often offers up a Dumpster fare superior to that of the linen tablecloth place with the celebrity chef.

When it comes to vegetable oil or biodiesel, water is not your friend. In a Dumpster, water causes vegetable oil to start to decompose, resulting in free fatty acids that get in the way of making fuel. When making biodiesel, having water in the reaction tends to produce soap instead of fuel. And, in the engine, water leads to rust and is an impediment to combustion.

If it is allowed to settle, water always finds its way to the bottom of the vegetable oil, so whether collecting oil from a five-gallon carboy or a Dumpster, the trick to avoiding water is preventing the pump intake from hitting the bottom of the container. The top layer of used fryer oil generally is where the chunks reside. Left over fry bits and trash that has been thrown into the Dumpster tend to float, so sucking from the middle is the best way to get the good stuff.

In general, pulling from Dumpsters is the hard way to go. The poorly paid staff who fill the Dumpsters typically are untrained and have a tendency to dump oily wash water into the mix. Retrieving the vegetable oil after it leaves the fryer and before it hits the Dumpster is the ultimate scenario and generally is easy to establish for small quantities. It is not hard to find a chef who is sympathetic to the cause and will pour the used oil back into the five-gallon plastic carboys from which it came.

On one occasion, when we were experimenting with ways to scale up our operation and I was growing weary of scooping and constant grease-runs, we tried our hand at purchasing used fryer oil. Doing so was an adventure in itself, and I published a blog entry about the experience:

Rendering Plant Shopping Trip

Today Scott Rowe and I journeyed to Fayetteville to buy some grease from a rendering facility. It was a veggie-oil shopping spree. What a trip.

We started the day by running over to Stacey's to tell the septic system guys to get working on the house she was buying. I asked the size of the tank on their tanker truck. The answer was 1,500 gallons. It's wild. Since taking delivery on the 336-gallon tank last Sunday, I've started eyeing cylindrical tanks with a sense of volume.

Down at Summer Shop we have 55-gallon drums. We have a 163-gallon grease-water separator. And the 336-gallon storage and dispensing tank. When you look at all three together you get a sense of gallons per space. On the drive to Fayetteville, Scott and I played "Guess the Volume" on various tanks, both stationary and mobile.

The rendering plant processes 16 million pounds of animal and vegetable byproducts a week. Trains of tanker cars roll by. Large numbers of 8,000-gallon tanker trucks come and go. Methane is flared off through a smoke stack that appears to have an apple juice can attached to it with baling wire.

In many ways it is like the scrapyard. In fact, it is a scrapyard. There are rows and aisles and groupings of similar merchandise. It's a mall of grease.

Shooting skyward in all directions are large, heated vessels, probably around 20,000 gallons each. Here's "yellow grease" from poultry rendering. Here's "brown grease" from bathroom grease traps. Here's feedstock from dead animals. Over there is the plant that processes only turkey guts. It's bizarre.

Scott was doing titrations on the hood. Straight turkey came out at the edge of what we can easily turn into fuel. We picked up 1,080 pounds of veggie that had been dried and centrifuged to free it of crud. We were told it had an MIU rating of 2 percent. MIU stands for moisture, impurities and unsaponifiables — that is, anything other than the grease itself.

We also picked up some straight turkey feedstock and some veggie and turkey feedstock to do mini-batches for the Alternative Fuels Garage up at the NC Solar Center. Rachel is collecting fuel samples from a wide variety of feedstocks to create a visually striking display.

At one point Scott and I were standing beneath a huge grease dispenser, waiting for a sample of grease to titrate. "What is your dream titration number?" I ask.

"I won't be able to get that," says Scott.

"Forget that, man. What is your dream number?"

"I can make fuel out of twelve. Ten would be good. Anything less than ten is a dream for me."

A spray of hot grease baptized Scott when one of the rendering plant workers reversed the direction of the flow without closing a key valve.

On the way home we were riding low. And he was soaked.

Sixteen million pounds a week ship through this little plant in Fayetteville. Unbelievable. To top it off, the grease was useless. Scott and Rachel tell me it is almost entirely water. Is there a "returns" counter at the rendering plant?

———————— ≋ ————————

It turns out we had the titration number wrong. When using virgin vegetable oil, the recipe calls for 3.5 grams of lye per liter of oil. When using waste oil, the recipe calls for 3.5 grams of lye plus X, where X equals the titration number. Scott's dream number should have been closer to five, not ten, which explains why we procured over 1,000 pounds of useless veggie. We ended up composting that load.

At Piedmont Biofuels we are blessed with a giant shopping mall that is too uptight to allow Dumpsters out back. The poor restaurateurs are required to tote the veggie out in five-gallon carboys. These carboys are handy at the outset but tend to pile up quickly and are exceedingly difficult to dispose of. Although they are high-density polyethylene (HDPE #2) plastic, they are oily and therefore contaminated from the recycler's perspective. Our county government does not want them. Our local plastics recycler will not take them. We once shipped a load to an outfit in Greensboro that said they would take them, but they ended up charging us a disposal fee.

It is easy to be overwhelmed with used five-gallon containers. Because they typically are closed on top, they are exceedingly hard to wash and dry. We have tried cutting them apart with carpet knives and power washing them. We have tried using degreaser to clean them. Whoever can invent a use for greasy carboys is sure to stake a claim in the future of backyard biodiesel.

We now have an arrangement whereby our veggie comes in from the high-end mall in 55-gallon steel drums, which we pump from the ground to a barrel on the back of our truck. Each drum weighs around 400 pounds, making them a materials-handling menace. Fifty-five gallons of used fryer oil is a lot. It's a heavy, slippery, stinky, dangerous mess. But 55 gallons is not a lot of fuel. That's not quite two full tanks for my 1992 Dodge pickup truck.

Once the vegetable oil is in hand, it needs to be filtered. Filtration is much faster and easier under pressure. Those who are running on straight vegetable oil need it to be filtered to about one micron, and there is a wide variety of ways to do this. The most common is a filter with a 12-volt pump powered by a car battery. The little diesel-fuel transfer pumps that commonly are used for this task are underpowered and tend to wear out and die. I find them loud, slow and annoying compared to a manual barrel pump, but this opinion seems to be in the minority at Piedmont Biofuels.

I once did a grease-run with a new co-op member who was espousing the virtues of 12-volt. I threw my barrel pump in, hoping to have a John Henry-like competition to see who could move more veggie faster. When I hit clogs or chunks with my barrel pump, I quickly reverse and continue on my way, whereas a clog in a 12-volt can cause the pump to smoke and die.

The competition never happened, since the two 55-gallon drums of veggie we had set out to retrieve were deep behind some Dumpsters and only the 12-volt had a hose long enough to reach. On other occasions when I have offered to take my barrel pump along I've been politely rebuffed, and sometimes we've had to come home without a full load because the silly little 12-volt pump gave out.

I'm not sure the "manual versus 12-volt" debate will ever be settled at Piedmont Biofuels, and while it rages on, the dead pumps accumulate behind the door in a pile of embodied energy we often neglect when calculating our energy balance. Here is part of a blog entry on the subject:

Manual Versus 12-Volt

I must confess there is a chink in the armor of Piedmont Biofuels. It is a hotly contested debate that does not appear to be subsiding. My position has been clear from the start: I love gravity, manual pumps and designs that depend on energy from humans. The same cannot be said for Leif and Rachel.

As a group of chemical plant designers who exist inside a great swirling conversation of invention, fantasy and innovation, we generally do OK. The debate appears when it comes to moving liquids. I like to barrel pump by hand. Leif and Rachel like to hook up little electric pumps with elaborate back-flush and filtration systems.

I like to sock-filter veggie, which involves hanging a fabric bag above a target vessel, filling it with grease and letting it slowly drip. To increase the rate of flow, simply add more grease to the filter, increasing the pressure on the sock. Leif and Rachel like to pump grease through automobile filters using 12-volt pumps hooked up to a battery. They argue that they catch both particulate and water with this method. I find the 12-volt gear loud, whiny, slow and subject to clogging.

The same disagreement arises when it comes to pumping biodiesel. I like a rotary hand pump affixed to a 55-gallon drum. Rachel and Leif will go to bizarre lengths to use 12-volt instead. I've seen Rachel maneuver her huge Dodge pickup into impossibly tight spots to get proximity to her battery so that she has juice.

If a weight is less than 60 pounds, I'm inclined to muscle it, especially if there are two of us around. Leif, on the other hand, will virtually replumb Alien Baby just to use its pump function, and it's a 110-volt device.

I don't think this schism is likely to break up the garage band. But it adds a persistent edge, which recently surfaced when I had a manic episode and outfitted Double Bubble with bungs.

"Bung" is a technical term for a threaded hole. As soon as you begin interacting with steel drums of almost any size, you will begin your relationship with bungs. Smaller eateries store their used veggie in metal drums.

Some bungs are plastic, some are steel and most have gaskets of one kind or another. Step one when entering the world of bung is to get a bung wrench. It's a tool that provides grip and leverage over large bung covers on one end and has a lever for small bung covers on the other. I didn't have this advice when I started out and I had to interact with bungs using vice grips, needle nose and snub pliers and screwdrivers backed by hammers, oxygen and acetylene. Save yourself a bunch of grief and buy a bung wrench. Whether you are collecting or storing SVO or finished fuel, the ability to successfully handle bungs is critical.

I suppose my initial resistance to entering the world of bung was that my previous familiarity with the subject came from human anatomy. I don't believe I'm alone in this. When the Library of Congress released the untold hours of President Johnson's taped telephone conversations, I remember hearing him complain that one particular pair of pants he owned had a tendency to "hike up into my bunghole."

When it comes to barrels, a bung is an interface. I like popping the cover and screwing in a metal collar that has an adjustable pin. I then drop one of my beloved barrel pumps into the drum and set the level to any height I want.

Last Friday I pulled in to get some fuel and was presented with the dilemma I have experienced since Double Bubble arrived. Ever since we started washing, I have begged those drawing the fuel off to slip some unwashed into a 55-gallon barrel on the side for my personal use. The barrel rests on a homemade cart, rolls effortlessly around and has a dedicated barrel pump affixed to its large bung.

But more often than not, I show up looking for fuel and am greeted by two huge plastic drums with the huge plywood tops Matt and I made at Chessworks, with a mix of washed fuel and goop inside. In order to get the fuel, you need to draw it off the top. Carefully. On this particular day, I found the user interface to Double

Bubble maddening as usual, and I recalled the two mangled metal drum lids Tuesday had brought home from the scrapyard. They would never seal a drum properly again, but their bungs were intact.

I grabbed the lids, tossed them on the truck and went to Chessworks. I fired up the plasma cutter, cut the bungs from the lids and carefully blew six holes around each one. I then drilled out two holes in the plywood so that the bung would lie flat. Taking a page out of the Greasel songbook, where they attach aluminum radiators to plastic fuel tanks, I affixed metal bungs to the plywood lids, thinking I could get a barrel pump in place atop the wash step and be able to roll off as much fuel as I needed.

When Leif and Rachel showed up I was delighted to show them my latest innovation. Leif was quietly bemused, looking at me as if to say, "Do you really have that much time on your hands?" Rachel was much more straightforward. I told her that once the collars I had ordered arrived, I could have barrel pumping capabilities on Double Bubble.

"That's awesome," she said, pointing to an errant 12-volt system that has long sat idle in the corner. "That one needs a bung to get started so we can start using 12-volt on these guys ...."

Filtering without the pressure of a pump is problematic. Gravity draining with sock filters hung over buckets is exceedingly slow. Heating the vegetable oil prior to filtration can have a dramatic impact on speed, but there also is a chance that waxes in the vegetable oil will liquefy, pass through the filter and reform in the fuel tank upon cooling.

The kings of straight vegetable oil filtration are Tom Sineath and Eric Henry over at T.S. Designs. They are running a homemade reactor on the back dock of their sustainable T-shirt business and have designed a solar-heated sock filtration system that can push hundreds of gallons through one micron in no time at all.

Eric is the consummate promoter and Tom is a thoughtful inventor. Together they have put together a successful textile business based on organic cotton and nontoxic dyes. Eric picked up his first methanol from me at Summer Shop, and Rachel and I took him a few more gallons for his second batch. His factory in Burlington is a remarkable place where active solar, sustainable landscaping, water reuse, fair trade coffee and all the principles of a reduced ecological footprint combine in the name of commerce. An Elsbett convert, Eric drives around on both straight vegetable oil and homemade biodiesel, and his enthusiasm is causing others to join the movement.

In the early days of our backyard explorations, we worried about this entrepreneur from a town less than an hour away overshadowing our efforts, but we were able to overcome our apprehension and provide him with free methanol and advice on getting started. Our openness has been repaid ten times over, as he has sent people to classes down at the College, referred countless others our way and shared his filtration expertise. Remarkably, T.S. Designs still collects veggie in five-gallon plastic carboys.

The relationship between Piedmont Biofuels and T.S. Designs was an early test of our ability to share and has been one of the many relationships that have led us to our "open source" philosophy. No secrets. Nothing proprietary. Help anyone with anything — even if it means helping others sell biodiesel in our backyard.

In backyard biodiesel the methods of collecting used veggie vary with the circumstances. Whether it's collected in five-gallon carboys or in 55-gallon drums or is delivered off the back of a pumper truck, used veggie is crucial to most backyard operations. Manual or 12-volt, sock filtration or pump, without ample quantities of good used veggie there can be no fuel.

seven

Feedstocks

Most people think of biodiesel as something that comes from soybeans, and this is largely because the forces of Big Soy have promoted the product. If you read the résumés of the board of directors of the National Biodiesel Board, it is readily apparent that Big Soy runs the NBB. Money from soy growers and the soy lobby was used to get the EPA's blessing for biodiesel.

Old-timers still refer to the product as "soy diesel," and soy makes its way onto bumper stickers and hats and other paraphernalia. Indeed, many of the big players in biodiesel — West Central Soy, Archer Daniels Midland, Cargill — are the "Who's Who" of Big Soy.

No one can blame them for creating a new product from their harvest, yet the preponderance of soy in the biodiesel industry means that alternative feedstocks commonly are overlooked or unknown.

Soy is, after all, a rather mediocre substance to use for making biodiesel. Fuel made from soy does have superior cold flow characteristics compared to fuel made from animal fats, but not as good as fuel made from Canadian canola. Soy-based biodiesel also has higher NOx emissions than fuel made from other feedstocks.

The oil-per-acre yield of soy is not remarkable compared to that of other crops. In fact, the only reason to use soy as a feedstock is the marketing of the industry behind it.

Let's face it. Soy is just a nasty little bean they have been trying for the past forty years to find a market for. You can dip soybeans in chocolate but they cannot compare to peanuts or almonds. You can dry them and salt them, but they are awful compared to potato chips. You can compress them into a gelatinous slab and flavor it six ways from Sunday, but tofu remains in a flavorless, joyless niche on the supermarket shelves. After forty years of sampling and rejecting soy products, we might as well just burn soy in our combustion chambers as we drive down the road.

It's worth noting that the soy oil from which biodiesel often is derived is a sidestream of the crushing process designed to make cake, or meal, that can be fed to animals. The oil is simply a bonus that could represent as little as ten percent of a crushing operation.

It always has struck me as odd that, despite soy's limitations, many of the new biodiesel plants on the books right now are going to be using virgin soy oil as their feedstock. In North Carolina, a state with a soy deficit — that is, it imports more soy to feed to its many pigs than it currently produces — there is a 30-million-gallon biodiesel factory based on virgin soy feedstock being planned. While it is the brainchild of the Soy Growers Association, the initial beans for the plant will be coming on a tanker from Brazil.

The "sustainability" part of biodiesel has been forgotten — which often is the case when it comes to the decision making of Big Soy. I have mixed emotions. Part of me feels it is better to ship a tanker of beans from Brazil to North Carolina than a tanker of crude oil from the Persian Gulf. After all, the beans are renewable, perhaps some farmers will get a better price for their crop and a straight soybean spill is somewhat less troublesome than an oil spill. And perhaps a 30-million-gallon virgin soy biodiesel plant in North Carolina would allow farmers to change over from tobacco or less viable crops and make a living on the land. It's hard to say.

Although soy is the popular feedstock, all we need to make biodiesel is fat. Beef tallow, used vegetable oils, peanut oil or liposuction remains all would constitute viable feedstocks. When I first had my toe in the legislative process and was actively lobbying the powers that be in North Carolina for a tax holiday for biodiesel, I published a blog entry that touched on some of these issues:

Feedstock Neutrality

There are two bills in the finance committee of the North Carolina legislature right now. HB 1705, sponsored by Joe Hackney, started out life calling for a simple tax holiday on biodiesel. I felt this approach was simple and easy and would have no opposition. It was suggested that the NC Department of Transportation (NCDOT) would object to the bill because it could mean a loss of revenue.

I thought that was nonsense, since government fleets (including NCDOT) are the predominant users of biodiesel in this state and government fleets don't pay taxes on their fuels. At the time the bill was proposed by Hackney, there was a single gas station in North Carolina pumping B20, with a couple of others pumping a B5 blend. That means the loss of revenue to NCDOT would be miniscule — unnoticeable, actually — and I was confident they would not object. But they did. And they replied to Hackney's office that they would drop their opposition if the tax holiday pertained only to biodiesel made from waste vegetable oil.

Joe seems to like this idea, since it precludes giving any breaks to Big Soy and would give a boost to small producers. I'm grateful the idea got as far as it did and I have spent a lot of time with his staffers speculating on how much waste vegetable oil there is in this state. I've heard a range from one to three gallons per person per year. If North Carolina has 12 million people, there is a bit of waste vegetable oil here. But while I am grateful, part of me resists the whole idea of specifying what biodiesel is made from.

On the other side of the House, Rep. Joe Tolson has introduced HB 1636, which also is in finance right now and is being read tomorrow. It started out life giving blenders' tax credits to those making biodiesel out of "pure virgin soy."

I hit the ceiling on this one and had to be talked down by my friends and colleagues in the movement. Since North Carolina has a soy deficit, specifying virgin soy feedstock was crazy. That part of the bill has been changed, and it apparently now defines biodiesel by the ASTM specification, which is feedstock neutral.

The two bills put me in conflict with myself. I'm not sure I can continue to support the tax holiday if it is limited to waste vegetable feedstocks. It leaves out too much potential energy. What about beef tallow? Or oil from black soldier flies? Mustard seed? In the front yard of the refinery there are beautiful swaths of yellow mustard seed. Oneas, who is rapidly becoming known as "Dr. Feedstock," is experimenting with a variety of non-soy virgin crops. What if one of them is a hit?

With regrets, I think it is critical that our government not legislate what biodiesel is made of. We should leave that to the producers. People who make the fuel can fight their own battles with free fatty acids and glycerin content. As long as they hit the ASTM specification they can call their product biodiesel and away we go. For consistency, we need to maintain feedstock neutrality in our legislation and focus our policy directives on how to give biodiesel a boost after it has been created.

One thing that attracts me to biodiesel is that it can be created from local resources. Let's use virgin soy where soy grows well and canola where the climate favors it. Let's use animal byproducts where they are in abundance, and anything else we can find. Let's use liposuction remains in Palm Springs and dead possums in Moncure and let's have our government regulate uses and taxing of the finished product and not meddle with feedstocks at all.

Let's stay feedstock neutral. As producers, all we need is a fat. Give us a glycerin molecule with a few carbon chains and we can make fuel.

One of the persistent questions from people new to biodiesel is: can we produce enough to meet all our fuel needs? I hate this question, not only because I find it hard to answer but also because it implies that if we can't meet 100 percent of our fuel needs with biodiesel we shouldn't bother making it at all.

The best stab at answering this question that I have seen comes from Michael Briggs at the University of New Hampshire. He takes a starting point of 138 billion gallons annually to power the US economy. To get that number, take the non-diesel energy consumption, convert it to more efficient diesel engines and then add current diesel use. Relying on algae as a feedstock, Briggs claims our fuel needs can be met with biodiesel.

There is plenty of fat in this country. Look around. There is fat running down the sides of Dumpsters in parking lots. Supermarket shelves are lined with fat. There is fat on every street corner. In fact, we have a fat epidemic.

A more interesting side of the feedstock debate is the ethical question of whether we should be using our valuable croplands to grow fuel. I once tackled this subject in the blog:

Food as Fuel

Today I did an interview with Bill Moore of EVWorld in Omaha, who has a site dedicated to alternative fuels and such. He was remarkably well informed and managed to take me onto some less familiar terrain. We talked about the biodiesel process, about backyard and small producers and about the NBB. And then we got into the ethics of burning food.

I tried to separate hunger, which is a distribution problem, from production. I learned about this a long time ago from Tony Kleese. I'm not sure what Tony's title was then. Perhaps he was running the Farm Tour or teaching sustainable agriculture over at the College or heading the Carolina Farm Stewardship Association. Whatever it was, Tony knows about food. He knows about its production and its distribution.

We were dining at the General Store Café one day and I was settling in to a Doug Lorie Famous New Mexican Green Chile Burrito. Tony explained that it is our food distribution system that is broken. People are throwing out perfectly good food on one side of the fence and on the other side people are hungry. This is true even if we don't grant Monsanto a patent on seed.

I had a more tongue-in-cheek go at the subject shortly after that:

GMO Monocropping

Tonight I was reading Kumar's *Fueled for Thought* and his entry inspired me. He recently had taken his mother to see Percy Schmeiser, a Canadian farmer who tackled Monsanto on some GMO seed corrupting his canola crop. Schmeiser was speaking as part of Mendocino County's Measure H initiative in which a California county banned GMO crops.

My limited understanding of the world's food supply is that we can produce more than enough food for everyone to eat but there is a disconnect between our production capabilities and our distribution capabilities. We have plenty of food and we have plenty of hungry people but we are not doing a good job of getting the food to those who need it.

When I lived in Marshalltown, Iowa, as a teenager, we would routinely stop by the fast-food joints right after closing time to intercept chow that was being toted from the kitchen to the Dumpster. It wasn't fine dining but it was still hot and was comparable to what earlier customers had paid for — and it was within our budget.

Now I prefer the window seat at Elaine's on Franklin, where the quail is generally just right (if you can think of it as gourmet rather than as a small, headless action figure). But through my work in biofuels I've encountered college students who routinely Dumpster dive for food. And I've encountered "freegans," which describes people who will eat anything that is served, as long as it is free.

Clearly the problem of hunger has nothing to do with production, which means GMO production must be tied to greed. If it's not about increasing production for the hungry, it must be about licensing and limiting and lining someone's designer pockets.

I have to say I admired Daryl Hannah's guts when she dropped the phrase "GMO monocropping" onstage at the National Biodiesel Board Conference. And I equally admired the NBB's Joe Jobe's response, in which he likened the topic to the "abortion debate," by which I presume he meant unproductive.

In his entry on *Fueled for Thought* Kumar suggests that creating biodiesel out of GMO would be a good way to get rid of it, and who can argue with that? Rather than letting it rot on tankers off the coast of Africa, where they won't take it to feed the starving masses, we ought to use it to feed our hungry combustion chambers.

Although most of our work at Piedmont Biofuels has been done with waste vegetable oil, we do have a USDA-registered research farm where Oneas does research in non-soy feedstocks. So far he has been working on sunflower seeds, mustard and oilseed radish.

Other possible feedstocks include fish oil, cottonseed oil, hemp oil and jatropha. Jatropha is a weed that grows in Asia and Africa and has an amazingly high oil yield. Today it grows bountifully in roadside ditches and is not widely cultivated for its oil.

And there is the black soldier fly, a possible feedstock that has held my imagination since I first encountered a mason jar of it at the Iowa State Energy Center. Researchers in North Carolina sprinkled black soldier fly maggots into hog waste. Apparently they consume a massive quantity of waste as they travel through their life cycle. At one point, they all begin to climb, bloated and fed, to a quiet place to enter their next stage.

Researchers built wooden ramps that the larvae could successfully ascend, only to fall, or "self harvest," into troughs beside the hog waste. Squishing them at this point in their lives produces rich, black oil that could be ideal as a feedstock for biodiesel.

I've never seen a black soldier fly in the wild. Apparently their adult stage lasts about two days, during which they simply breed and die. Without any feeding requirements as adults, they have no cause to interact with humans. The current research on the black solider fly is being spearheaded by Tom Matthews at Washington State University, and he is connected in some way with Jon Van Gerpen in Idaho.

The question of whether we can grow enough biomass to power our economy reminds me of a conversation I have had several times with my brother Glen. He is in the business of shipping electrons created from the wind. Wind is his feedstock. In the normal course of his renewable energy business, he does outreach and education and he routinely is asked something like: "But what do you do when the wind doesn't blow?"

I haven't heard his answer up close, but I am assuming he does not say: "Right. True. Sometimes the wind does not blow. Therefore we should not bother developing this free, renewable, non-polluting, locally made resource that comes to us from the sky."

At Piedmont Biofuels we also do a terrific amount of outreach and education and we routinely are asked: "But can you grow enough to create fuel for all of us?"

To this perhaps I should say: "I don't know. Maybe we cannot grow enough. Therefore we should not bother developing this free, renewable, less-polluting, locally made resource that comes to us from local Dumpsters."

The similarities are striking.

There is a persistent feeling that we have not even scratched the surface of biodiesel feedstocks. And I am not underestimating the American farmer. Those who say we cannot possibly grow enough fuel to power our economy might look back at World War II, when this country figured out how to produce everything from guns to airplanes to ammunition on such an unprecedented scale that we still have excess manufacturing capacity fifty years later.

When biodiesel moves into the mainstream as our next renewable fuel, I will not be astonished by the way American agriculture can ramp up production of vegetable fats.

eight

Birth of a Co-op

P iedmont Biofuels is incorporated as a cooperative, as defined by the laws of North Carolina. There are a number of reasons we are a co-op. The first is Leif, who is a co-op enthusiast. When I met him he lived in co-op housing. When we needed a domain name, he insisted that I buy a dot-coop. I think he likes the egalitarian aspect of co-op structure, which to me seems much fuzzier than the fuel itself.

Under our incorporation, we can make a profit. We are like co-op grocery stores or feed stores that are free to behave in a profitable fashion. Because of our association with the College, which already had a tax-deductible foundation that handles donations and grants, we did not feel the need to incorporate as a charitable undertaking.

However, while we have long been free to make a profit, the motive to do so has been so far down our lists that it does not happen reliably. Because we lacked the profit motive, lots of people have joined the co-op and lots of people have contributed to our success. We routinely stage "refinery Sundays" that attract volunteers from far and wide, and thanks to our co-op members and volunteers we arguably are farther ahead than we would be as a simple corporation. I've recounted several "volunteer days" in the blog:

Refinery Sunday

A couple of weeks ago, Rachel, Scott, Oneas and I put in a long day at the refinery. The day culminated with a tired group and a box of Rolling Rock on Oneas's front porch. It felt great, but we speculated on what it would be like if we had a larger group of people working on the project.

And so today we had "Refinery Sunday." Ten people showed up to volunteer. It was amazing. Scott spent the day hammering nails on a hot tin roof. There is a shed behind the refinery that we call the "Princess Palace" after Rachel, who once claimed it as her new residence. That shed is scheduled to become the Biodiesel Bed and Breakfast, and a new roof is a great start.

Murat and Cynthia ripped half the roof off the cement slab and started constructing a "walk-in closet" that we will be able to use for both methanol delivery and 55-gallon grease drums. From the beginning we have wanted a place to enclose our methanol storage and at long last it is underway.

Leif and a bunch of people successfully made a 65-gallon batch. Phil built a battery storage shelf next to the 336-gallon elevated tank in the sideyard. Ward designed the new plumbing and did the carpentry required to let us "meter" the gallons that fly by. Gravity is not enough for successful metering, so we are using a 12-volt system that will be powered by a solar panel on the roof.

It was a crazy day. Hot and "management intensive." As one of the hosts, I worked hard furnishing guidance and tools and logistical help to the crowd, and my brain hurt. At the end of the day my hands were cut and scraped and I was thoroughly sore and burnt out. It felt great. I walked through the woods and dove off the rope swing into the pond, which was full of children and dogs and the usual summer Sunday crowd.

The refinery got a huge "bump" today. Leif pulled in with the newly painted, newly loaded tanker and sold some fuel off the truck for the first time in history.

It was a remarkable day at the refinery — a volunteer ballet. I still have a hard time realizing that a roof was torn down, a wall was put up, a batch of fuel was made, a new roof was put on and a 12-volt system for pumping fuel was designed and is almost in place. The work that went on today normally would take a month.

—————————————— ≈≥ ——————————————

And while it is true that dozens of volunteers are unlikely to show up to help a corporation increase its profits, it also is true that hiring the work done would perhaps be more efficient. The absence of a profit motive means that is not an option for Piedmont Biofuels.

Another important aspect of the co-op structure is that it significantly reduces the costs of erecting self-serve fueling stations. If you want to install a B100 pump that is open to the public, it needs to have a double-walled tank that is tied to the grid, surrounded by concrete car barriers, lighted to code and regulated by the USDA's weights and measures division. A good starting point is $15,000. Yet if you are a private fleet, simply trying to fuel your trucks, you can have an elevated single-walled fuel tank with a nozzle on it. I could easily put one together for free.

The co-op is closer to the "private fleet" side of the street. After all, fuel is available only to co-op members, not the public, and we can tell the fire marshal that our members are trained on how to fill up their vehicles.

Training, by the way, involves the following instructions:

1. Pop door to gas tank.

2. Unscrew lid.

3. Insert nozzle.

4. Squeeze handle.

5. Stop when tank is full or when auto-shutoff nozzle stops.

6. Wipe spilled biodiesel off paint job (it will act as a solvent and eat the paint away).

Another reason we are a co-op is that the authorities do not allow us to sell our homemade fuel. It is perfectly legal to make your own biodiesel, put it in your car and drive on the road with it. The main fuel-governing authority, the EPA, allows this. The IRS still wants to tax every gallon, and rules that anything put in a fuel tank is taxable. They have forms to file that allow users to record the number of gallons they use and remit the tax. In North Carolina, the state taxation authorities did not know where to start on homemade biodiesel as long as no transaction occurs. Their feeling has been that if there is no transaction there is nothing to tax. Just before 2005 they changed their mind on that, and they now have the same disposition as the IRS. Their ruling is a massive setback for biodiesel in our state.

The worker members of Piedmont Biofuels sign a membership agreement, pay $50 a year and commit to working five hours a week. Some haul grease. Some make fuel. Some do carpentry. Some work on websites. Because their worker contribution can be seen as "making their own fuel," they are able to put homemade fuel in their tanks and drive onroad.

The co-op structure is exceedingly handy as a way to move some homemade fuel and stay legal. According to our membership agreement, members can make their own arrangements with the IRS, and they do pay for the fuel-making supplies (methanol and potassium) they consume. Supply costs hover around a dollar a gallon for homemade fuel.

When it comes to store-bought fuel, which we buy from commercial biodiesel manufacturers and distribute, there is no clear reason to be a co-op. We sell the fuel to anyone who wants it, whether they have a membership or not, and we sell it all for the same price.

For users of store-bought fuel, the only benefit of being a member of the co-op is that we allow members to serve themselves off the tank at the refinery while we typically meet non-members and fill them up. Because the tank at the refinery is based on the honor system — fill up, drop your money in the lock box, drive away — we like to know who is using it and seem to trust members more than non-members.

The co-op structure does mean that we get membership fees as a revenue stream, and it is true that some people who don't even drive diesel vehicles — people who have no use for either homemade or store-bought diesel — have joined our co-op.

My only other experience with a co-op is with Chatham Marketplace, a whole-foods co-op that a bunch of people, including my wife, are trying to open in Pittsboro. They have spent years on membership drives, have a one-time fee and will offer discounts on food once they open.

Both Mom and Dad are active in co-ops at this point, and it is confusing to our two young sons. Mom has hundreds of members, no location and no sales. Dad couldn't care less about having members but moves some fuel. The situation has not yet led to co-op on co-op violence in our household, but it has been interesting watching the two co-ops unfold.

Our participation in the world of co-ops has spilled over to other groups. Our membership agreement, which is on our website, has been used verbatim by other biodiesel co-op startups. One of my favorites is down in Asheville, and I once wrote about it in the blog:

Giants of Asheville

At the Biofuels to Biomass conference a young guy named Brian stopped by the booth and seemed keen and excited. I don't know how tall he is, but I have no choice but to look up if I want to make eye contact with him. The problem with enthusiasm on the show floor is that you seldom see it bear fruit. It's not that it doesn't bear fruit. It's just that you generally are not around to see this happen. And I think after having hundreds and hundreds of conversations that seldom lead to visible results you become indifferent to the wild ideas of others.

I'm not talking about the "I'm going to get a bus" crowd. That would include the many people I have talked to who are about to sacrifice a portion of their young lives to travel around the country on a biodiesel powered bus. To evangelize. And spread love. And help the movement along.

Brian was not one of these. He was intense. Focused. And he was getting a co-op off the ground in Asheville. Hardened by too much booth duty, I discounted him as part of the "I'm going to start a co-op" crowd. But at the Southeast Energy Expo Brian introduced me to one of the founders of the Asheville co-op, Solon (named for the ancient philosopher and lawgiver). He was as tall as Brian and he blew me away. Throw in a third guy, Mac, who knew Rachel from a past life and is just as tall as the other two.

Rachel, Leif and I broken up an intense meeting of Piedmont Biofuels and headed out for beers at a favorite watering hole where, in a heads-down conclave amid the madness of the bar, Brian, Solon, Leif and I got down to it. We hit on the implosion of the Biofuels mailing list. At the time, girl Mark had exited the list and barbs were flying back and forth between her and Keith Addison. As one poster put it, "When Mom and Dad break up, the kids suffer." We were the kids. We delved into safety, structures and legalities. They spoke of offroad and bioheat markets, and explained what they knew of the Asheville demographic. They have critical mass. They have momentum. They are forging into the foggy frontier of what girl Mark calls "the small producer," operations like ours that could turn out 50,000 gallons a year "in buckets" and don't.

I was jazzed by their intensity. We spoke of recipes and reactors, of policy and politics. We touched on a micronodal model of energy production and on the philosophy of the grassroots movement. We talked about greed and about public relations, and we were astonished when Solon kept pace with us on the June 5th contract price of a barrel of crude.

They need some space. And some structure. And some investment. But they will cross over from the backyard to the small producer and beyond. They are going to be giant.

———————————— ≫ ————————————

When I am giving refinery tours, I sometimes will use the phrase "welcome to co-op structure," usually with some derision, when I cannot find a tool or a pump is left on or someone has closed up shop without cleaning up after a shift. When the electric company disconnects power because the unpaid bill (which is in Rachel's name) applies to the property that Scott owns, which is used by the co-op to make fuel, all the neighbors lose water, as the refinery shares a well with surrounding houses. When I take their calls, I apologize profusely and I am quick to blame our co-op structure.

Because we are a co-op with worker members, we are populated by volunteers and have the same problem as churches and homeowner associations and every other volunteer group on the planet, which is that a small core group does most of the work.

People find their way to biodiesel for a wide variety of reasons. Whatever draws them, all types populate the movement and by and large a co-op is viewed as the most egalitarian and least greedy of the structures available to meet their needs.

nine

The Shadow and
Grassroots Conferences

In the winter of 2003, Piedmont Biofuels was established as a co-op, making some fuel and doing a whole bunch of outreach and education. Rachel was loosely connected with others in the movement, notably Marty Stenflo of Boulder Biodiesel. I had a vague idea that we were part of a larger whole, but knew very little about what was going on in biodiesel in America. Rachel and Leif were all over Keith Addison's amazing website, Journey to Forever, and Veggie Avenger and Joshua Tickell's Veggie Van (now Biodiesel America) site. I was happy to leave the virtual layer of our exploits to them, since I had left the sculpture business and taken a job with an Internet company. Spending my days online, I preferred working on the physical side of biodiesel.

But I was teaching a class on energy down at the College and was spitting out blog entries with fair regularity. Girl Mark had discovered the blog, and Piedmont Biofuels had a lot of traction. We were planning on attending the first-ever National Biodiesel Board (NBB) Conference in Palm Springs, and girl Mark announced that she was going to stage a California B100 users' conference in Claremont the week before NBB. To everyone's astonishment, including our own, Leif, Rachel and I decided to attend. By the time girl Mark had commitments from Marty and Lorance and the folks from Colorado, her conference had suddenly swelled beyond the boundaries of California and become known as the Shadow Conference. A small auditorium was procured with the help of Kaleb, a chemist, activist and enthusiast who had connections in Claremont, and activists from all over poured in.

Joshua Tickell was there with a Hummer, touting the virtues of hydrogen, with some folks from a commune in the desert. Kent Bullard, who runs a National Park Service fleet off the coast of California and was buying huge quantities of biodiesel, was there. Kumar and his wife Sunny of Yokayo Biofuels were there.

It was an astonishing group. Our tickets were taken at the door by Nicole Cousino, the creator of *Fat of the Land,* a remarkable video about a group of women who travel across America making biodiesel out of waste veggie along the way. Her efforts predated the famous voyages of Josh Tickell, and there was a decided sentiment in the air that she deserved the fame and glory for the popularization of biodiesel.

Betty Biodiesel was there; she apparently entertained crowds in Iowa. Jennifer Radtke was there, as was her business partner, Sara Hope Smith. They run the legendary Biofuel Oasis in Berkeley. And Kimber Holmes was there. She coined the bumpersticker "Biofuels are Peaceful." I attended her heartfelt discussion of how she opened a B100 pump with virtually no money and no margin in the fuel. She was powered by pure passion and was proud of her permitting victories, through which she managed to get biodiesel mentioned in the local code.

Charris and Nick from Grassolean were there. The first is hysterical, the second is brilliant and they want to disassociate the word "diesel" from the fuel. Diesel connotes black smoke in the minds of the American public, and they believe Grassolean is a better name. Daryl Hannah was there in her El Camino with a series of fuel containers strapped to the bed.

I was a babe in the woods at the Shadow Conference, but I forged some deep relationships with other activists that I cherish to this day. The work of Piedmont Biofuels was completely colored by those we engaged with at the Shadow Conference. It was where the seed of our tank truck was planted. The conference steeled our resolve on the way to NBB. And it still informs our thinking on biodiesel.

Many left the Shadow Conference and headed through the valley to Palm Springs, showing up at NBB wearing simple B100 stickers on their shirts and lapels. The trip itself was remarkable and I wrote about it in the blog:

Fouling Our Nest

Our day started early. We left our Claremont hotel dump in the dark of the night and took Rachel to the airport. I wished she could have stayed for the whole trip, but that sentiment was overridden by the stress of getting to Palm Springs.

Today the groundhog in Pennsylvania saw his shadow, and I worried that it might be a bad omen as we headed for NBB. We were eastbound on Highway 10 when we saw the astounding wind farm in the San Gorgonio Pass. I had a hard time concentrating on traffic amid the hundreds of turbines on both sides.

As the static began to overtake Inland Radio, we saw the most amazing sunrise over the Coachella Valley. And in the distance the San Jacinto Mountains were obscured by smog. No doubt the sunrise had its beautiful colors enhanced by particulate and fluorocarbons and NOx and SOx. In Thomas Berry's introduction to Anna Caldara's *Endangered Environments* — a picture book — he described the Los Angeles area's smog during sunrise. His remarkable description and the adjacent photograph were powerful enough, but seeing it firsthand was overwhelming.

Breathing became hard for me. I was sneezing and wheezing and felt as if I had spent the night sleeping with a cat. Leif commented more than once on the stench of our drive, as the air recirculation on our rental car did not work.

At 6 a.m. on a Monday, the highway was packed. It is hard to imagine how people could watch their views deteriorate without modifying their behavior. It was difficult not to compare the majesty of the windmills to the presence of the smog and not to wonder: will all that non-polluting energy from wind be too little too late?

———————————— ≫ ————————————

At the end of January 2005, the NBB held its second conference in Fort Lauderdale, Florida. In a fit of over-enthusiasm, girl Mark and Rachel decided that Piedmont Biofuels would host a conference ahead of time in Pittsboro, which could be followed by a trip to NBB.

I was opposed to this idea. Our feedstock supplies were in peril, we were on the cusp of starting our industrial project, a girl Mark reactor design-build workshop had trashed the co-op and I was overwhelmed by the work that lay before us. Rachel was not fazed by all this and forged ahead with conference plans, so like any well-disciplined member of a garage band I choked off my objections, stepped to the back of the stage and started delivering the bass lines she needed.

My biggest contribution was suggesting that we take the train from Raleigh to Fort Lauderdale. I had a romantic vision of having Tami and the boys on a once-in-a-life-time train ride up and down the eastern seaboard. I supplemented the vision with the discovery of a Curtis Mayfield song, "People Get Ready," that talks about boarding a train and hearing the "diesels running." Arlo and I were learning it on the piano together and it immediately became the theme of the Pittsboro Grassroots Conference.

Tami brought her formidable marketing and design talents to the table. She created conference folders, buttons and coffee mugs. She also decided that a new career as movie producer would hit the spot, and arranged to have the entire conference videotaped and recorded. I remember the night at the kitchen table when we were batting button ideas about. Rachel wanted to leave the Piedmont Biofuels logo off the buttons. We knew they would have to say B100, in homage to the Shadow Conference's stickers, and we knew the buttons would surface at NBB and help identify the true believers. We came up with "B100 Community," with our web address in the fine print, and no sooner had we coined the term than an e-mail came in from the NBB's Joe Jobe, concerned about our impending conference and calling us a "faction." Leif and Rachel ducked the reply, and I started a productive exchange with Joe in which I described us as being "more like a community than a faction."

I ended up assembling the lineup for Friday night, and we packed the multipurpose room at the College. Larry Shirley, the director of our State Energy Office, had agreed to speak and he electrified the group with a speech on the state of renewables in North Carolina. What impressed me most was that he waded directly into climate change and global warming, unheard of at conferences in these parts. Most energy conferences and most renewable energy projects are funded in part by the federal Department of Energy. Indeed, much of what funds Larry's operation comes from DOE. And it seems that wherever DOE monies are present, talk of global warming is absent. Larry bucked the trend for one glorious night and had the crowd spellbound with his own energy and enthusiasm for biodiesel.

I did a brief talk after Larry's and published it in the blog:

Grassroots Movement

I got an e-mail from Joe Jobe the other day. He's the executive director of the National Biodiesel Board. He has been following this conference from afar — he said he wished he could be joining us on the train ride to Florida — and was reflecting on some of the issues that have arisen between the NBB and grassroots activists in the past.

The last time a bunch of grassroots activists showed up at the annual NBB convention, they were wearing some laser-printed stickers that said B100. They were pushing for increased fuel quality. They wanted the NBB to open-source their human health effects data. They wanted a voice for small producers.

The NBB responded. They put together a small producers' working group and they came out with a small producers' membership that was half-price. They tabled discussions of quality policing until another time. From their perspective, they listened to what the grassroots community had to say, and instead of getting a "thank you" they were vilified for their failure to pass quality regulations.

Something you have to know is that the NBB has a devil of a time speaking for its members. Its members are diverse, with their own agendas and their own axes to grind, and some of them no doubt would love to live in a world without grassroots agitation.

And something the NBB has to know is that no one speaks for our membership. We are diverse, with our own agendas and our own axes to grind, and some of us would love to live in a world in which we could sell fuel unfettered by the NBB.

The reality is that the NBB has to put up with us and we benefit from its existence. We are strange bedfellows but we are bedfellows just the same. I would love to give you some perky, upbeat message about how great it would be if we could all just get along, but it's not that simple. Let me read a dark passage from Kumar's blog, *Fueled for Thought:*

On one side, there are giant petroleum/chemical/ag corps that produce poor quality fuel consistently and have created a national lobby that exclusively focuses on incentives, repeatedly quashing attempts at real understanding of sustainability and quality control. On the other side, we have a sea of enthusiastic progressives, 99 percent of whom are unaware of the ramifications of small-scale production, legal, safety, environmental, and otherwise.

Sure, Kumar is oversimplifying the positions of both sides, but in doing so he makes the point succinctly. We don't need to agree with the NBB. Those in the room who share enthusiasms for both Linux and biodiesel probably never will agree with the NBB's monopoly on access to proprietary data. But you know what? I argued that point with Joe Jobe in Palm Springs and he listened. Genuinely listened.

We can be critical without being hostile. We can be appreciative when things go our way. And though we may not speak with one voice, we do not need biodiesel to deteriorate into factions. The NBB knows we are here. They know we are the ones in the classrooms. They know we evangelize the fuel. They may have members who think we are impossible to approach or deal with, but as an organization the NBB is ready to talk and to listen to our ideas.

What is the grassroots biodiesel movement? You're looking at it. It's in this room tonight. We are in the backyard and we are in buckets and we are the ones driving around on this fuel. We write the blogs and we live on the lists and we know our biodiesel. The grass-roots biodiesel movement is supportive — supportive of one another and supportive of biodiesel.

Shortly after my talk the room rearranged itself into a circle and a hundred people introduced themselves and explained why they were present. Ten different states were represented. There were co-ops that were starting, co-ops that were thriving and co-ops rising from the ashes of ones that had gone before. There were farmers, curious newcomers, entrepreneurs on the scent of opportunity and seasoned biodiesel activists who knew the whole scene.

For me a critical part of the evening was watching those from the renewables establishment — people like Larry Shirley and Anne Tazewell — looking out over a conference that was standing room only. Piedmont Biofuels was no longer a lone voice for B100. I felt we had moved from being extremists on the wacko fringe to being simply one part of a giant community that shared our vision for the fuel. It was remarkable and edifying and an excellent launch to the conference.

There are no motels in Pittsboro, North Carolina. We packed the General Store Café on Friday night. Bed and breakfasts around the county were filled. The refinery slept ten people. Tami and I put up several. Out-of-town guests were billeted all over the place, and the larger community's acceptance and assistance were touching. People who don't even own diesel vehicles and are not part of the movement were taking in boarders for free.

Saturday the conference ran like clockwork. Rachel kicked it off on time and sessions began. There was a panel on algae as a feedstock and one on emissions that received rave reviews. Girl Mark and the Flexistentialist crowd from St. Louis held a discussion on their collaborative writing project, which has gone on to become a wiki and could emerge as a definitive text for biodiesel.

John Bonitz, operating the Piedmont Biofuels tank truck in the parking lot, pumped off hundreds of gallons of fuel. Sara Hope from Berkeley was there, weary from her travels, contributing to a workshop on small biodiesel businesses. Rob Del Bueno, the rock star turned SVO enthusiast, was there from Atlanta, holding court in one of the College's automotive bays. Charris from Grassolean in Colorado stood in for a rap session after lunch.

Our tours of both the refinery and our proposed industrial project were cold and wet, and the National Guard closed the highway in front of the College because of impending ice. The weather caused the conference to dissolve in the late afternoon — Southerners retreat at the mere mention of snow or ice — and to this day Rachel longs for better "closure."

About fifteen of us piled into the Amtrak station in Raleigh. Tired, running on adrenaline and buoyed by the afterglow of a remarkable conference, we embarked on a nineteen-hour conversation that would lead us to Fort Lauderdale and the NBB.

ten

The National Biodiesel Board

The National Biodiesel Board is not a bad organization. In some ways it is too easy a target. Despite all the barbs that are directed against it, it has a legitimate value.

At the heart of the matter is their proprietary use of human health effects testing. This is an expensive battery of tests and an exhaustive literature review necessary for any fuel to pass muster with the EPA. The NBB claims to have spent three million dollars collecting the human health effects data and having their research accepted by the EPA. This money came from the soybean farmers of America from what is called the soy "check off" program.

Biodiesel poses a problem for the EPA because their procedure calls for hooking rats up to pure emissions and calculating how long it takes them to die. In the case of biodiesel, the rats don't die. They can live long, productive lives in a climate of pure biodiesel exhaust. The NBB testing involved ten-week-old rats that spent six hours a day, five days a week for ninety days in the presence of biodiesel exhaust before being carved up and scoured for evidence of damage. One of the great ironies of the biodiesel industry is that the NBB's fuel of choice for the environmental testing was B100. Somewhere between their stellar testing results and today, the term "B100" has slipped from the NBB's vocabulary.

Legend has it that two organizations took a run at the EPA at the same time. Shaine Tyson of the National Renewable Energy Lab (NREL) led one effort. Apparently their intention was to "open source" the testing data so that anyone producing biodiesel would have legal cover from the EPA. The other effort was spearheaded by the NBB. Theories of conspiracy abound, rumors of political intervention from powerful agribusiness interests are rampant, but at the end of the day the NREL bid was turned down by the EPA (it seems they had used the wrong engine for their tests) and the NBB's bid won acceptance.

Anyone who wants to sell onroad biodiesel needs to produce the data for the EPA, and the only way to do that is to join the NBB. It's unfortunate that most small producers of biodiesel arrive at the NBB's doorstep for this reason. Membership for small producers used to be $5,000 a year and was dropped to $2,500 at the NBB meeting in Washington, DC, in the summer of 2004. Interestingly, by the time of their annual conference in January 2005, almost ten percent of their membership had subscribed in the "small producer" category. The creation of a "small producers' working group" and the reduction of fees were the direct result of girl Mark's Shadow Conference and considerable grassroots agitation. People like Kumar from Yokayo, Charris from Grassolean and Kent Bullard from the National Parks Service guided the effort from the NBB side.

Unfortunately, the NBB is characterized first and foremost by old-fashioned proprietary beliefs with a monopoly on human health effects data. It is true that their vocabulary begins with B2 and ends at B20, and I have sat in audiences and heard NBB employees say, "You can run on B100 but it will wreck your engine," which does not endear them to the B100 community. It also is true that their web presence is lacking and they do not have nearly the coverage on the lists that they should. They can routinely be savaged up one side and down the other and not even know it.

Yet when I'm in Washington, DC, talking to my elected officials about impending biodiesel legislation, I follow the NBB line. My own brief brushes with the legislative process have taught me it is better to be big and well-organized than to be a radical voice in the wilderness, even if that means speaking for the value of blends or helping to get a petroleum blenders' tax credit on the books.

The NBB funds valuable research, and some of their contractors, like Steve Howell of Mark IV Consulting, are the most articulate and informed people in the country on biodiesel and the diesel industry.

When doing outreach and support for biodiesel, it is very nice to have the NBB in your corner. Piedmont Biofuels may not be able to change the mind of an engine manufacturer like, say, Caterpillar, but Caterpillar *will* listen to the NBB. And when the customer is caught between the Piedmont Biofuels version and a conservative engine manufacturer's version, sometimes it is the NBB's version that tips the scale in our favor.

Yes, the NBB represents the interests of Big Soy. And yes, they have a monopoly on health effects testing. And yes, they speak largely for blends. And yes, they are critical to the impending success of biodiesel.

The relationship between the grassroots movement and the NBB is fraught with tension. Grassroots activists tend to be anti-greed and anti-GMO and to look down their noses at blends. NBB members are overwhelmingly interested in shareholder value and return on investment and view the grassroots side of the movement as an annoyance — occasionally making claims that backyard brewers cannot possibly meet the quality specifications or are using dangerous practices because they are unable to invest in safety.

Yet time and again it is the small producer or the backyarder who ends up doing the technical support and outreach for the commercial producer's fuel. It simply is not unusual for small producers to have a better handle on quality.

And at the same time, backyarders are by and large the sales force for biodiesel. When a municipal fleet starts including a two percent blend in its mix, it's not nearly as interesting to the public as the person who makes fuel in the backyard. People's imaginations are ignited by the veggie Dumpster, not by an ADM balance sheet, which is why commercial biodiesel needs to cultivate its relationships with backyard brewers and small producers.

The National Biodiesel Board should be funding analytics labs at colleges in every state. They should encourage home-brewing. Instead of staging golf tournaments, they should stage seminars on reactor design and biodiesel safety. I made this observation to a reporter from *Wired* magazine at the National Biodiesel Board Conference in Fort Lauderdale.

I was covering the conference in the blog, and my number of unique readers hit an all-time high:

Day One of NBB

There are 1,000 people at this year's National Biodiesel Board Conference. My guess is that they expected around 750, which is why the line to get a beer at last night's reception was so long that the reception was over before the beers were served. Many went without lunch today. The place is crushed, and while there is plenty of grumbling, who cares? The more the merrier.

I missed most of the morning's general session (train lag, I suppose) and got there just in time to hear Charris and Daryl Hannah knock out a rap tune and to see Daryl give Neil Young an award on behalf of the NBB for his use of biodiesel. Once again, Daryl took free rein at the microphone, and she not only wore one of our B100 Community buttons on stage, she also gave backyarders everywhere a plug, mentioned the B100 community and told the audience of soy producers to avoid GMO monocropping! She was remarkable. Just like last year. An inspiration.

By all accounts the first session of the Technical Track, which was Biodiesel Fleet Experience and OEMs (original engine manufacturers), was a monumental waste of time. Apparently there were no engine manufacturers present and everyone I met who attended was disappointed.

I went on the Policy/Regulations Track to an Update on National and State Legislation, and it was fantastic. Scott Hughes, the regulatory director of the NBB, was very good and I gained some vocabulary that I have been lacking. All my efforts at shaping the legislative landscape in North Carolina have been on the "demand side" — that is to say, for a tax holiday for the fuel. And yet our state's renewables establishment crowd does all its work on the "supply side" — for things like tax credits for petroleum blenders. Realizing this difference was exceedingly helpful.

Last night at a poolside reception, Anne Tazewell agreed to take a run at NCDOT with me to see if we can get them to drop their opposition to my tax holiday idea. And today in the hallway, Tobin Fried, our Clean Cities coordinator, objected when I referred to North Carolina as a "biofuels backwater."

I came out of the morning session and did an interview with *Wired* magazine. They had twigged to the differences between the "corn-and-milk-fed crowd" (the reporter's term) and the grassroots folks at the conference.

In the afternoon I jumped into the Users Track to attend a session on Quality. It too was fantastic. George Kopittke of Griffin Industries kicked it off with a wonderful description of quality as being that which is "fit for purpose." He went into the technical differences between a Bic pen and Cross pen and likened it all to biodiesel. He impressed me.

There was a flash of tension in the Quality session when some guy from the US Postal Service complained about some off-spec "homemade fuel" he had bought. Straightening him out, so that it was clear he could not possibly have been buying from a home-brewer but rather from an NBB member, was a little touchy.

I felt that Gene Gebolys from World Energy stole the show in the Quality discussions, but since we buy from World Energy I am predisposed in his favor. He made the point that biodiesel is not a product but a service, and that our customers expect flawless performance from us every time.

Instead of pretending that biodiesel cannot be made effectively in the backyard, the NBB should embrace the notion that individuals can meet their own fuel needs. The NBB has the mistaken belief that their power comes from their membership — a small, elite group of large companies — when in fact they could swell their rolls and increase their power by embracing the backyarder.

They have an old-fashioned publicity machine. They send out press releases on how some petroleum blender in Missouri got an award for pumping a million gallons of B2. No one cares. The buzz for biodiesel comes from the newspaper story on "Fueled by French Fries," which is not sent out by some publicity flack.

Just as it is easy for reactor designers to sit around and throw darts at the work of others, it is way too easy to criticize the National Biodiesel Board. The NBB needs the grassroots activists to keep the buzz alive on the fuel and to continue the evangelizing, outreach and education we perform and we need the NBB to represent our interests.

eleven

Two Professors

Largely because of the work of Jon Van Gerpen (now at the University of Idaho), I think the heart of biodiesel research in America really belongs to Iowa State University and to the Iowa State Energy Center in Nevada, Iowa. Greg Pahl, in his wonderful book, *Biodiesel: Growing a New Energy Economy*, credits Charles Petersen of the University of Idaho as one of the leading pioneers in this field. Another early promoter was Thomas Read at the Colorado School of Mines, who often gets mentioned as the critical inventor of small-scale backyard processors. However, I think that today Van Gerpen's program at Iowa State is where backyard and industry meet.

Van Gerpen is a quiet, curious and unassuming man who leaves an indelible impression on everyone lucky enough to attend his courses. Though a creature of the academy, with strong ties to commercial biodiesel, he seems to secretly enjoy and respect the grassroots aspect of the biodiesel movement.

The Iowa State Energy Center is a remarkable place, funded in part from a tax on each Iowan's electric bill. The center has a complete biodiesel factory in one corner, an ethanol project in another and a gasifier in another. Those who take their seats to study biodiesel in Iowa sit down next to folks from Malaysia, Singapore, Toronto and Japan. It literally is where the world comes to study biodiesel production technology, commercial-scale biodiesel business management and biodiesel analytics.

During my brief stint there, I was seated next to investment bankers from New York, PhD chemists from rendering plant operations in the south and venture capitalists with chemical engineering degrees. As a back-yarder, I stuck out like a sore thumb. And while the chemistry quickly left me behind, I was the only one in the class — including the instructors — who actually had driven on biodiesel. It astonished me to see everyone show up in pickup trucks and SUVs to teach biodiesel at the world's foremost academic institution in the field.

I was delighted to have dinner with Van Gerpen in Harrisonburg the night before Virginia's first-ever biodiesel conference, and he confessed that student pressure over the years (including mine) had forced him into a diesel. It was remarkable to hear him ask

Rachel for some advice on how to maintain his car. Since prior to his work in biodiesel his academic career was based on diesel technology, to have him finally driving around in a diesel struck me as moral congruence.

But Van Gerpen doesn't do morals. Or agendas. He simply does science. One night over beers in Iowa he explained how as a scientist he provides objective data for industry, governments and activists alike. I conceded a grudging respect for his position, for while it leaves him silent on the geopolitical aspects of oil and biodiesel, it allows him to present the fact that soy feedstocks have higher oxides of nitrides in their emissions than other sources, and his scientific integrity allows him to do so even though the research may be funded by Big Soy.

I was pleased to learn that Van Gerpen was an occasional reader of Energy Blog. He told me he had saved the meeting minutes, resolutions and notes from all the discussions on biodiesel he had participated in since 1992, thinking the material would one day make a good book since the biodiesel industry in America was so new. I imagined the proceedings of the ASTM working group where a meaningless petroleum distillation point was included in the biodiesel specification. Van Gerpen would have been at the table for those discussions. I was crestfallen to learn that he had thrown everything away when he moved to Idaho.

At the NBB meeting in Fort Lauderdale in 2005, I offered Van Gerpen a B100 Community button, but he declined respectfully to attach it to his lapel. "It is political at all?" he wondered.

On the fringe of biodiesel research is Alex Hobbs of North Carolina State University. He runs the NC Solar Center there and has forgotten more about BTUs than most will ever know. My first contact with him came when he offered a supportive telephone call in the winter of 2001. He knew about funding. He encouraged backyard production. With his southern drawl, his kinfolk's familiarity with stills and his vast knowledge of both mechanical and chemical engineering, he struck me as an encouraging nutcase.

Over a year later he drove his massive SUV out to Summer Shop and poked through the piles of our reactor designs gone by. He seemed as comfortable with a mason jar and a 55-gallon drum as he would be advising the governor on energy policy. Hobbs ordered a reactor from us that became known as 1.4. Leif did most of the work on it, down at Summer Shop, and I provided technical and logistical support. It currently resides at the Alternative Fuels Garage at the NC Solar Center.

With his blend of folksy stories, his prodigious technical know-how and his somewhat Santaesque white hair and beard, Hobbs has been a terrific contributor to the movement in North Carolina. He is an accomplished public speaker and an inspiration to his students, and his assistance to us has been everything from cash to having us build a reactor for him to encouraging us to persevere in the face of disappointments we have encountered along the way. My favorite line from his speeches is:"Why do we keep going to the devil to get our fuel when we can grow it right here on God's green Earth?"

I also have seen him put up a slide of Earth's atmosphere and indicate with a laser pointer,"There may be a problem putting our waste right here," which is the closest reference to global warming from officialdom I have heard at the handful of Department of Energy conferences I have attended.

When I learned he was bringing a class of students down to the refinery, I wrote about it in the blog:

Dr. Hobbs Comes to Town

Last Thursday night I got word that Alex Hobbs, the guy who runs the NC Solar Center, was bringing a vanload of graduate students for a tour of our new refinery.

Yikes. At the time we were piled high with grease. Leif recently had taken a Sawzall and cut a hole through the wall of the doublewide trailer that adjoins the back porch. We were basically a mix of sawdust and grease, wedged around our new — undeveloped — wash stations, giant yellow vessels on barrel carts that Leif scored for a few bucks at a surplus auction.

I managed to pull in Matt, a local carpenter who is as good as he is finicky about the work he chooses. He spent Friday framing in a door so that our hole in the wall would appear to be an office. Rachel worked into Friday night, sweeping and unifying and setting things right. And I managed to get the entire place spotless Saturday morning. It was very cool working to the gurgle of our new drain-back solar thermal installation. Once we add a manifold, we will be heating not only the facility but also our feedstock with solar hot water.

We began the tour at the General Store Café, with fourteen mechanical and chemical engineering students plus Eric and a new guy from T.S. Designs. We packed the place. I watched Vance and company scramble to dish out the orders and I can't help but be reminded how a busload of students who have traveled from Raleigh will provide a momentary shot in the arm to the General Store's income statement.

At lunch it is all NBB. We are updating Hobbs on the Shadow Conference, on the small-producer debate and on the EPA registration issues. We talk about the next NBB meeting in DC in July and, as always, he is thinking ahead of us. I beseech him to help us get the North Carolina legislators to attend the NBB meeting. He thinks that is a fine idea and asks, "What is the one message you want to take to them?" I have no idea. As far as I have gotten is that if we started working now we could pack the place with our delegation. Crafting a single most important message will take some work.

During the refinery tour I make it clear that Piedmont Biofuels has a large number of positions open right now:

- We have an immediate opening for a full-time lobbyist — someone who can get biodiesel on both the North Carolina and the federal radar screens.

- We need to open our research and development arm, and require a full-time researcher who will do nothing but experiment with our process. Suggestions include ethanol, Magnesol and premixed methoxide.

- We have an immediate opening for a full-time grant writer — someone who can eat whatever they kill. Hobbs brightened up at that one and may put some of his formidable resources to work for us.

- We have an immediate opening for a plant engineer. The successful applicant will move us forward in areas such as diesel electrical generation, construction of a solar still for methanol recovery and perfection of our wash process.

- And we have an opening for someone who can lead our analytics effort. We need to acquire about $70,000 worth of gear (see grant writer) to assemble a quality lab.

I think I caught some of their imaginations, right up to the point where I made it clear that none of these positions is paid ....

Throughout it all, Hobbs tossed out insights for the group. He reflected on his time with Procter and Gamble here, or on a grant proposal he wrote there. As I pointed out design flaws in Alien Baby, he directed us to the food processing industry that has "one-step" cleaning requirements built in.

We had a good tour; I thought we showed very well in our new facility. Having T.S. Designs present gave me a chance to plug the micronodal energy production model I love to harp on. They are less than an hour from us, making fuel to power their small group.

If there were enough of us, perhaps we could upend the oppressive "top down" infrastructure that controls our current energy regime. When it comes to fuel, we generally are at the mercy of others. If you didn't have a coal mine or a railroad, you couldn't really play in yesterday's world, which was dominated by coal. And if you don't have a pipeline, an oil well or a refinery, you can't really play in today's world of petroleum. The nice thing about biofuels is that you can build your own refinery in your garage. If everyone did that we would have a shot at changing our typical energy infrastructure to a micronodal system that might be borderline sustainable.

And if we are going to pull that off, we need help from people like Van Gerpen and Hobbs who understand that backyarders are critical to the development of the fuel, to its standards and uses and to the biodiesel industry itself.

twelve

The Renewables Establishment

It is hard not to compare the current actors in biodiesel to those of the solar movement, and indeed many of the people who have been preaching renewable energy for decades are now hitching their wagons to biofuels because that's where the money is. The North Carolina Solar Center, for instance, has opened an Alternative Fuels Garage. At the ribbon cutting (where we heard Kim Kristoff on stage with regulators and politicians say, "Your government is dumb as a post when it comes to renewable energy") it felt as if those of us standing around the burn barrel on winter nights hoping for separation were like the first solar installers thirty years ago.

A huge difference between then and now is the Internet, and I'm repeatedly discouraged by the aging renewables crowd's relationship to the net. As grassroots activists, we are completely networked. Discoveries spread instantly; breakthroughs are old news in the flash of a few keystrokes. Most biodiesel information begins on the message boards, blogs and mailing lists online, and yet this is a world that is largely undiscovered by the renewables establishment. The great democratization of the Internet does not seem to apply to those old warriors who have been in the renewables sector for the past thirty years.

I was lamenting this fact one night at Raleigh's Museum of Natural History with our Clean Cities coordinator, complaining that those with money and influence over policy seldom are current on biodiesel issues because of their failure to use and understand the Internet. The next day I received an e-mail offering a series of webcasts that the national Clean Cities folks had prepared. Their idea is that you sit down at your computer at the appointed time, call in on the telephone and watch them share their PowerPoint presentation online. The entire concept flies in the face of the net.

The reason the Internet got traction is that it allows users to access the content they want when they want it in the form they want it. As a medium the net introduced "point casting" as an alternative to "broadcasting." The broadcasting paradigm throws content out at a specific time in a specific way and we are forced to consume it "as is." The Internet freed us. It is irritating when folks like Clean Cities decide to impose a broadcast mentality over a point-cast medium to claim a stake in the online world.

George Bush just signed biodiesel tax credits into law. As the news spread, there was a dire need for interpretation. What would happen to the price of fuel? Would we end up with overpriced inventory? When would the effect be seen on the street? What would be the time frame from the president's pen to the pump?

While the renewables establishment was fumbling around in an attempt to send out "talking points" to help those of us on the front line answer consumer questions, Kumar published his dad's interpretation in his blog, *Fueled for Thought*. Kumar is a committed distributor of B100 in Northern California who is working on assembling his own production facility. He is an industry pundit whose opinions have tremendous weight. He has been in biodiesel for a long time. He is thoughtful and exceedingly knowledgeable. His dad is a CPA who understands the tax code.

On Kumar's blog, those online who travel the net as a routine part of every day found an instant and sound interpretation of the tax credits. Most of the renewables establishment probably never had read a blog. The fact is that subscribers to the Biofuels list and people who read girl Mark's prolific posts typically are better informed than the current renewables establishment.

The National Biodiesel Board's site is powered by "Grassroots," but unlike true grassroots efforts it seldom works properly. After you submit information you generally get an error message and need to call and give the same information over the phone. Their idea of e-commerce is to have you fill out all the information online, print the form and mail it in with a check. Someone at Big Soy (which powers the National Biodiesel Board) understands the need for a website but they don't quite understand what it could be used for. When their e-commerce section is "Closed for Repairs," they often have a flashing banner advertising their information technology services.

The same is true of the rendering world. While it may be a stretch to refer to the rendering industry as part of the renewables establishment, a good example of my complaint can be found at <www.renderers.org>. Anyone wanting to purchase some yellow grease for a commercial biodiesel operation might start at this website, but since the members are listed alphabetically it would be prudent to know all the area codes of North America prior to conducting a search.

The digital divide between activists and the renewables establishment is frustrating indeed. It leads to conversations in which the informed backyarder has to bring the technocrat up-to-date before the discussion can progress to the next level. If the information on a website is stale or inaccurate, the user can simply move to another point. If the information in someone's brain is out-of-date, it is a chore to bring him or her along.

Technologically challenged though they may be, the renewables establishment controls the "free money" available in biodiesel today — and there is lots of it. Of course, there really is no such thing as free money. If I had a nickel for every time I've heard someone utter "You could get a grant for that," I could be retired by now. Just the same, I have taken a stab at writing grant applications and I have addressed this in the blog:

Our Friend Grant

About a year ago, Rachel came by with about two pounds of paper she had printed out down at the College. It was the application for a grant from the Department of Energy. I had written some grant applications in my past life, so I agreed to give it a read. I thought I would try on "grant writer" as a self-image, just to see if it fit.

I wrote the application and we sent it off to Southeast Regional SARE — Sustainable Agriculture Research and Education. What we wanted to do was build a "closed loop" down at the College, where we would plant an oilseed crop, press the oil, make biodiesel and run the diesel tractor. We figured twenty grand would do it. We made the first cut, but it got so weird and complicated that it wound up in a lateral pass to Leif, who ran with it for a while before deciding to punt.

And I wrote a grant proposal for the Department of Energy. It was so complex and convoluted that I swore I would never do it again. I'd rather file tax returns for fun. Rachel has attempted to soften my relationship to the process by referring to our application as "Grant," as if it is a child or a friend.

"Have you heard from Grant lately?"

"What's up with Grant?"

Almost a year has gone by since we filed our application for Grant. It's to build a mobile biodiesel processor that we can drag around the state to do demonstrations with. Whatever dollar amount the DOE gives us will reflect pennies on the hour for the energy we already have exerted on this project.

Grant passed first reading. We must have got the title pages stapled on right. Grant made it into the final twenty participants. Yesterday Rachel and I went down to Raleigh for the opening of the bids. It was tense and exciting — like watching a kid graduate from school.

A landfill needs $5.5 million for a methane recovery project. Appalachian State University needs $163,000 for a bookmobile on green energy. There goes that company that used to be Duke Solar. They need some cash for a feasibility study. And here comes our proposal. Wow. We are the low bidder. Only $52,000 to build a mobile biodiesel processor, staff it and tote it around the state for a year.

The envelope please ....

I would love it if Grant were to come our way. The money would come to Central Carolina Community College and give the biofuels program some credibility and traction. And it would get a mobile processor built that would travel around the state.

Now that we have made the cut, I'm pro-Grant.

But the other day Rachel showed up with another manila envelope.

"We should write a grant proposal," she said. I glared at her.

"You are going to hate me," she said, as she handed me the four pounds of paper. I suppose I will wade into it and see what the possibilities are, but I have to say that I don't see myself as a grant writer. I would much rather design a new filtration station or pass along the gossip of biodiesel than hang around with our friend Grant.

We ended up landing that grant, and Rachel and Leif have done a mountain of work administering it. We took the funds and built a sweet little mobile processor that currently tours the state, and we have the renewables establishment to thank for that opportunity. I covered the processor's completion in the blog:

Maiden Voyage

Tonight Zafer and I went down to the College to help put the finishing touches on the mobile biodiesel processor. Rachel was burning the midnight oil, doing a brake job on her Dodge to get ready to haul the unit to Asheville. This is the unit that was funded by "our friend Grant" from the State Energy Office and the Department of Energy.

We have been collecting pieces and parts for this for over a year. Lee Iron and Metal donated a double-jacketed stainless reactor vessel. John Bonitz at Celebrity Dairy pointed the way to "sanitary fittings" that attach to it. Rachel found a generator at Delk's Army Surplus over in Asheboro and I fished a trailer out of the woods down in Goldston. Mark Stinson had some leftover paint that he used on the trailer restoration. It can be described only as "Eye Popping Yellow."

Tuesday and Bob attached the generator to the trailer and modified the steel so that all the doors on the generator would open. Tuesday also prepped a piece of shiny scrap aluminum and Stacey Emerick knocked out a marvelous series of logos, including those for the Department of Energy, the State Energy Office and Central Carolina Community College. George drilled out sixteen holes for the generator's feet and Carpenter Matt wired up an electrical service from an old temporary pole that I had found in the woods.

Earlier today Rachel and I went to Chessworks to blow a pair of holes in her homemade trailer hitch. We ran out of oxygen in the middle of the project and took refuge at Kenny's Bus Farm to finish the job. Ken hooked us up with oxygen and acetylene so we could modify the receiver on Rachel's truck.

Leif has been the point man on design, plumbing and fitting and bolting his brains out. Because the monster generator is water cooled, we are taking the waste heat from the electrical generation and surrounding the reaction with hot coolant — similar to an SVO system but much more elegant.

We designed this last summer down at the river, and I remember fighting over whether or not we should add methanol recovery. I was overruled, and Leif has added the most amazing methanol recovery system made of copper pipe. I think Mac, one of the giants of Asheville, might have assembled the core of the system when he was in town for the girl Mark workshop. This is reactor 1.6, affectionately known as "Clean Tech," and anyone who sees it on the road tomorrow or visits it in Asheville is guaranteed to be impressed.

The Carolina Farm Stewardship Association (CFSA) meeting is being held this weekend in Asheville. Tami and the boys are going on behalf of Chatham Marketplace. I'm headed to the Outer Banks for a meeting about Piedmont Biofuels Industrial — our latest fantasy, which involves recycling an abandoned chemical plant into something that can produce biodiesel.

As part of the CFSA conference, Leif and Rachel will be demonstrating how to make biodiesel, live on the back of a trailer. The conference is sold out as usual and promises to be a fantastic weekend. The website for the Clean Technology Demonstration Project is <www.biofuels.coop/cleantech>. Tonight Rachel is wiring brake lights, affixing a license plate and working into the night. Zafer and I mounted Stacey's sign on the back ramp with four bolts — a fairly straightforward drilling project that took one thing off the list.

Everyone pulled together to hit the CFSA deadline, which has made for a crazy two weeks, but at the end of the day this combined effort is going to help us continue to get biodiesel into the consciousness of North Carolina.

It would be hard to describe this project as emanating from the renewables establishment, even though they were the funding source. In fact, it was just the opposite. Instead of involving long meetings around vast conference tables discussing renewable energy strategies, the project drew on the skill sets and capabilities of a wide variety of individuals. Building Clean Tech was a study in grassroots collaboration.

Tomorrow is her maiden voyage and I can't wait to hear how she fares ....

I have had many positive interactions with our State Energy Office, although they too have not fully grasped the Internet. I once invited the director to town to give him a tour of Piedmont Biofuels. His secretary asked for directions. I told her they were on the web. She asked me to print them and fax them to her so she could attach them to his day planner.

Technology aside, I was once summoned to their office and I wrote about it in the blog:

State Energy Office

Yesterday morning I fought my way from Chatham County to deep inside Raleigh's beltline for an 8:30

meeting at our State Energy Office. I must confess I was expecting opulence. I've been to parties at the governor's mansion and to meetings at the Department of Public Instruction (which educators refer to as the "Pink Palace"). I've checked in with the guards at the Department of Administration and I've frequented most of the museums downtown. All my brushes with the government of North Carolina have been characterized by largeness, grandeur and power. Until now.

The State Energy Office is supposed to be championing conservation, renewables and alternative fuels. They invited me down to talk about my legislative concept of "No taxes on biodiesel. Period." The office is located in nondescript flex space in a little industrial mall that's hard to find on Map Quest. I pulled up and parked in front of a badly painted sign for the North Carolina Boxing Commission, which maybe shares the building or the parking or something.

And I have to say I was impressed. Lights were turned off in offices that were not in use. Employees switched their lights off on exiting their offices. There were no Styrofoam cups in the darkened break room. Sugar for coffee was sitting in a glass bowl. I carry my own coffee cup and was delighted to see everyone else doing the same. They appeared to have a complete office recycling program. The parking lot was full of compact cars and hybrids and it was clear to me that the people of the State Energy Office mean it. They are conscious of energy. They get it.

Five of them gathered around a Spartan conference table and peppered me with questions. They knew their stuff. They were not a bunch of bureaucratic half-wits humoring some biodiesel wacko from Chatham County. They hit hard on increased NOx emissions. They were well- versed on ethanol (stumped me) and compressed natural gas. They were intimate with the fuel distribution lobby and the car dealership lobby and they knew the legislature very well. I got the distinct feeling they also knew biodiesel.

At one point, I talked about how our reactor was on display at Alternative Fuel Odyssey Day at Wake Tech. I said, "Some kid was examining it on one of the breaks, so I went over to explain how it worked and we were immediately surrounded by forty people."

Larry Shirley, director of the State Energy Office, nodded and said, "That kid was my son." I was blown away. These people are genuine. They are dragging their teenagers to alternative fuel conferences. My teenagers barely read the blog. These people are amazing. I don't know if they will line up behind my idea of "No taxes on biodiesel. Period," but they certainly entertained my views.

Going to the State Energy Office was a fascinating brush with the people who do the work for the State of North Carolina. Technocrats, bureaucrats, civil servants, public servants, whatever you want to call them, they are immersed in the energy field and are practicing what they preach.

I've done some speaking at their conferences. They are the ones who sponsored the NC Biomass Conference and Efficient NC. But these conferences are staged in the grandeur of NC State University and catered by contractors who don't even know what is being preached. While I find the conferences valuable, they remind me of the board meetings for the sustainability program at Central Carolina Community College, where everyone is offered drinks in Styrofoam cups and out-of-season fruit on Styrofoam plates.

I really enjoyed my trip to the State Energy Office. I get the feeling they are on our side and I am pleased to get all the help we can.

One of the best organizations in the renewables establishment is the Clean Cities part of the Department of Energy. I am forever singing their praises and clashing with them at the same time. They funded a tank distribution project for Piedmont Biofuels, the grant administration of which kept us from seeing the money for so long it about took us under. We had to get a loan to prop us up while we waited for the funds to flow our way. Just the same, Clean Cities is fighting the good fight and doing a good job. I covered one of their speeches in the blog:

American Fuels

One of the speakers at the recent NBB conference in Palm Springs was Shelley Launey, the national head of the Clean Cities Coalition. I should say at the outset that I love Clean Cities. In a world where the EPA is cumbersome, distant and confused and the Department of Energy has most of its head buried in the sand of the Middle East, Clean Cities has managed to fund the coalition movement. My guess is that Clean Cities is just a rounding number in DOE's budget.

Clean Cities is run on a shoestring. They work tirelessly on building relationships that support the replacement of fossil fuels. When our local coordinator wanted to go to DC to a conference, she had to stay with someone she knew. When she brings speakers to the Triangle area they sometimes stay at her house. Since we don't really pay the people of the Clean Cities movement, I try to pay them with compliments, and I did this at NBB.

Shelley Launey was a refreshing break from the "golf jokes" told from the NBB podium. She had statistics about the estimated gallons of fossil fuels her organization had replaced. She had an update from Washington. And she introduced the concept of "American fuels." At the time I liked her logic. She said people don't like the word "alternative." "We don't like alternative education or alternative lifestyles so let's change the name from alternative fuels to American fuels."

It was catchy. And couched in an inspirational speech. But the more I thought about it, the more preposterous it became. If you trace the lineage of Piedmont Biofuels, you start with Joshua Tickell, who refers back to his time as a student in Germany. I don't think Rudolph Diesel was from Missouri.

The problem with "American fuels" is that it smacks of the parochial, head-in-the-sand Bush administration's protectionist/isolationist/unilateralist sentiments that got french toast (invented by an American) renamed "freedom toast" by Congress.

Tonight Tami and I took the boys to The Greatest Show on Earth. Ringling Brothers/Barnum and Bailey were playing Raleigh, and we piled in. Chinese acrobats performed the opening number. The tightrope walker was allegedly making his American debut. I couldn't discern the accent on the horse trainer, who had his stallions rear up and bow to the adoring audience, but trust me when I say that his was not an Iowan drawl.

We must not allow "American fuels" to get traction or we will look as silly as those people at Pittsboro Little League who are still trying to sell "freedom fries." I'm not opposed to changing the vocabulary. Language is supposed to be a living thing. I started out liking Launey's revision to the lexicon, but time has weighed in against her.

I don't mean to find fault with Shelley Launey. She transformed the general session of that day's NBB conference from a waste of time to something worthwhile. And she has a tough job. When the Bush administration noticed that the Clean Cities Coalitions were doing excellent work, it forced them to recognize B2 as an alternative fuel. Even worse, since hydrogen-powered transportation is the Bush administration's solution for our clean fuel needs, Clean Cities now must include hydrogen in its conversations.

———————————— ∽ ————————————

There is an inevitable confluence of the B100 Community and the renewables establishment. We may meet through money, since backyarders will be tagged for their expertise as grant money flows into biodiesel. We may meet through evangelism, since backyarders are needed to make the case for biodiesel either from the podium or in the parking lot, where we constantly are asked to "show off" our vehicles. We may meet in the classroom or at the Rotary Club, since the NBB has just funded a Clean Cities request for a speakers' bureau that actually pays people to run around talking about biodiesel in North Carolina. We may meet in the policy layer, where seasoned veterans of the solar wars can guide backyarders' efforts. Best of all, we may meet over a tank full of biodiesel, since many in the renewables establishment are intrigued by our renewable fuel.

 *thirteen*

The Lure of the Producer

It is almost impossible to be involved in biodiesel without feeling the urge to "scale up" production. The moment a large batch separates and you know you have broken the blender barrier is the moment your imagination begins to run wild. If ten gallons, why not a hundred? If a hundred, why not a hundred thousand? If a hundred thousand, why not a million?

I'm not sure whom to thank for this instinct. Henry Ford, perhaps? There is something about the human animal that immediately makes the leap from "a little is good" to "more must be way better." I'm not sure if it's greed or ego or an innate desire to harness nature, but it's there, and it's there in almost every backyard brewer and small producer I have met.

When we first broke the blender barrier I was reading Jeremy Rifkin's *The Hydrogen Economy.* It's a disappointing book on a lot of levels, but his description of a micronodal model of energy distribution that has thrown off the yoke of the public utilities or Big Oil or the coal industry has been a profound inspiration to me. So with Rifkin to gird us, we successfully resisted the desire to scale up and devoted ourselves to perfecting small-batch processors that could meet a small number of drivers' needs using local feedstocks.

Indeed, one of the ongoing projects at Piedmont Biofuels today is what we call a "closed loop." It is largely Oneas's domain, but at its heart is the question of how many acres of what would a small producer require to grow oilseed, crush it and react it onsite and make enough fuel to power a small operation — a tractor and a market truck.

We have intentionally limited our thinking to small scale and resisted the desire to become commercial producers. That is, until recently.

In the summer of 2004 there appeared a moment when I thought I might have to move my brother's software company to a new facility. This came as a shock to me, since I had constructed his current building in 2000 to meet our unique needs so that we would never have to move again. I hate moving, especially moving businesses.

At that time, when I had been scrambling about for a place to build, I had encountered an abandoned industrial complex on the edge of Pittsboro. It was a strange-looking affair. It had big water and big sewer, yet it was in the middle of the woods. It had a guard-house and was surrounded by barbed wire. Apparently it was a throwback to the Cold War, and some cagey contractor had managed to convince government of the need to finance a complex in the middle of nowhere.

I walked through the empty buildings looking with the eyes of a sculptor, and my only thought was, "That's a lot of metal to cut out and scrap just to have an empty box to put a racquetball court in." It was too expensive for what we needed and it didn't fit perfectly, so I passed on the place.

Almost five years later, jarred by the news that the family software business might have to move, I revisited the place. This time, seeing the same buildings with the eyes of a home-brewer, I discovered an abandoned million-gallon biodiesel plant. I stared at the huge steel mezzanine and was stunned. It was as if the Cosmic Waiter had tapped me on the shoulder and said, "Sir, your table is this way. Come have a seat."

I started referring to it as "The Compound." It is four buildings on three and a half acres inside a fence. "Compound" is an accurate description, but those who visited it thought that "Compound" was too "Branch Davidian" and that it needed another name. In keeping with the do-it-yourself axiom of biodiesel, we needed to invent some vocabulary to handle this place.

I started calling it "The Park," which I thought had a nice ring to it. But immediately it started getting confused with Research Triangle Park, which is nearby and which we are forced to frequent from time to time. "Park" wasn't working, so we tried "Campus," but that lasted barely a week before people were getting it confused with the College. And so it remains a nameless place. Tami discovered that in the forties, before there was a Cold War to warrant such a white elephant, there was a cut-flower farm onsite that was known as "Sunshine Gardens." I think Tami is the only one that works for. As Mary Beth, one of our co-op members, remarked, "That's way too 'Care Bearsesque' for me."

When girl Mark learned of the existence of this place she reportedly grinned and said knowingly, "Ah, the lure of the producer." My guess is she has seen it a hundred times. There is Kumar of Yokayo Biofuels in Ukiah, strapped to the mast of his ship, telling his earplug-wearing crew to row onward, distributing ever-increasing amounts of B100 to a loyal customer base, while he alone is tortured by the siren's song of production. My guess is girl Mark has felt the tug herself a thousand times and yet somehow has been able to resist. Not me. I stand in the presence of an abandoned chemical plant and realize it is time to quit my day job and go into commercial biodiesel.

Before completely jumping off the dock, I invited Sam Moore from Burlington Chemical to town to take a look at the place. He's a third generation chemist with factories in Burlington. He drives over in his Honda Insight and pokes around. He equates making biodiesel with making wine and tells me there are two hundred chemical plants between here and Charlotte that could work. He also tells me that the site is superb and that the leftover gear lying around is worth more than the property itself. With his encouragement, the seeds were planted for what now is known as Piedmont Biofuels Industrial, LLC.

In biodiesel there is a general rule of thumb, promulgated in part by Big Soy and the National Biodiesel Board, that there is about one dollar of capital cost associated with each gallon of annual production. That is, to erect a million-gallon plant, have a million dollars handy. Some of the turnkey operations, like Biox or Capital Technologies, Inc. (CTI), come in slightly higher at around $1.20 a gallon. These are proprietary black-box affairs in which feedstock goes in one end and fuel comes out the other. The typical footprint of one of these plants is equivalent to that of an eighteen-wheeler. They are completely automated, usually with a proprietary catalyst, and all the buyer has to do is roll a plant off the back of a truck and voilà — they have a biodiesel factory.

I had explored both Biox and CTI from the backyard perspective, and I found they had similarities. Biox is funded by Canadian government money and predicated on a "new secret idea" from some University of Toronto bigwig, and part of their sales pitch is that if you don't adopt their technology now, you surely will be left behind. Never mind that when I spoke to them in the fall of 2002 they had yet to ship a working plant and that, although I frequently passed through Oakville, Ontario, in my travels, a tour of their facility was impossible.

Today Biox is planning to open a 60-million-liter plant, which represents more than half of Canada's planned production capacity.

CTI, which offers a business plan that has capital costs hovering around $1.20 a gallon, is a spin-off of Carnegie Mellon University — again, proprietary catalyst folks with a "better mousetrap." Although I dismissed them out of hand, I was encouraged by Anne Tazewell to dig deeper. My brother Glen was traveling through Pittsburgh in his Honda Insight the day CTI's e-mail started circulating in North Carolina. In the fall of 2004 he arranged a tour and found a company that was powered by venture capital, had one order on the books from a university in Brazil and had yet to ship a plant. While they are less arrogant than Biox, CTI's message seems to say, "Adopt us or die. We are about to revolutionize the industry."

With our open-source disposition, we were emotionally disinclined toward either process. More to my liking was the work Russ Teall had been doing at Biodiesel Industries. His model is closer to $1 a gallon, and he has several plants up and working in the real world. Pacific Biodiesel in Hawaii and Superior in Minnesota can make similar claims.

I was surprised when I heard Vivek claim he could complete a working biodiesel plant for $0.20 a gallon. Vivek was my lab partner in Iowa who threw away his job as a New York City investment banker to open a nine-million-gallon capacity biodiesel plant in Ohio.

Just when it looked like he was going to pull it off —
they had recycled an abandoned methanol distributor-
ship into a working biodiesel plant — their partnership
was torn asunder and the only fuel that came off the
line went into a boiler. Still, $0.20 a gallon caught my
imagination.

And then there was an outfit over in Tennessee that
in the spring of 2004 was selling biodiesel for $0.09 a
gallon less than petroleum diesel. I first heard of them
through Eric Henry, who had the idea that it was some
fat cat who was buying down the price himself and
swallowing the difference.

When I talked to them they claimed that there was
no buy-down afoot and that they were simply using
recycled chemical plants. They were amused by the $1
a gallon industry standard I tossed around.

Armed with encouragement from Sam, and bold
claims by Vivek somewhat supported by the folks in
Tennessee, I figured we could get this project going for
something less than $1 a gallon.

I had seen Kumar's business plan and had watched
him from afar as he shopped for the latest in biodiesel
plant technology from Austria and Germany. He has
chronicled some of his trials in his weblog, *Fueled For
Thought,* and I am not sure he has figured out the
financial piece yet. I also have watched Tom Leue's
fundraising efforts. He is sort of the elder statesman of
backyard biodiesel, whose story is well known to users
of the Internet.

Although Piedmont Biofuels always had focused on small-scale production, I'd had a constant eye on commercial concerns. Seven months before discovering the abandoned chemical plant, I published a blog entry on the subject:

Return on Investment

A few weeks ago there was a minor thread on the Biofuels list about "investing in biofuels." Some guy in the Midwest reared his head and asked everyone to send him a thousand bucks so he could raise $300,000 to build a biodiesel plant from scratch. It's a tough list, and the guy was savaged, went away and has not been heard from since. But there were a few echoes to his post, largely bemoaning the absence of investment in biofuels.

There is no shortage of investment opportunities. ADM, Cargill and Monsanto are easy to track and buy, and I have no doubt they will profit from biofuels in the years to come. The problem is that to many they represent incomprehensible greed that should never see our investment dollars.

Investing on the sustainability side of the street is much more difficult. Kumar and Sunny at Yokayo Biofuels in Northern California have a business plan that calls for investment. I've heard that Tom Leue at Homestead has one in circulation as well. Both of these would represent

direct investment in small businesses and both investments would be highly dependent on the success of
the founders.

There's nothing wrong with that, especially since
both Kumar and Tom know a lot about biodiesel production and distribution. But it is like investing in the
corner café, and not everyone is ready for that level of
risk. I put my investment dollars in the backyard and
down at Central Carolina Community College, and if
you count driving down the road on homemade fuel as
a "return," I've had a terrific return on investment.

I can't think of return on investment without conjur-
ing up images of Sky Generation, my brother's company
that hung the moon on a 1.8-megawatt wind turbine in
Ferndale, Ontario. Glen is the president of the Canadian
Wind Energy Association and he has forgotten more
about energy than most will ever learn. He's an MBA
with a keen sense of numbers.

His investment in wind, a project that seems almost
like his life's work, is paying a massive amount in intan-
gible dividends such as education and exposure to
alternative energy. It's created ecotourism and demon-
strated the existence of a previously undiscovered energy
source on Georgian Bay. None of these returns fattens
his bank account and, frankly, he was planning a return
for his investors that would stimulate investment in
more turbines.

We are still waiting. What saddens the investors in
the Ferndale wind turbine is not the absence of cash

dividends but the regulatory climate — the successive Ontario governments that have no clue about renewables. We also are saddened by the poor sales of Green Tags in Ontario. If more consumers participated in this program, there would be a larger pool paying a premium for green electrons and big wind investments would offer earlier returns.

And I'm afraid that investment in biodiesel is much the same. You can buy a Jetta TDI and join your local biodiesel co-op. That's an investment that will help change the world but its return won't help your daughter pay for college.

You can plunk down investment dollars in biodiesel startups, but that's not for the risk-averse. Surely investment in biofuels cannot lose over the next decade (wind either, for that matter), but if you want a return right now, you probably are best off buying a slice of World Energy. You will find them under the ticker symbol for "Gulf Oil."

———————————— ∼ ————————————

I had enough money to buy the compound and some knowledge of the market and business. I decided to skip the fundraising step entirely. I figured it could come later. Instead, I went to Leif and Rachel and asked for permission to use the Piedmont Biofuels name for a new limited liability corporation, Piedmont Biofuels Industrial, LLC. They granted permission and agreed to let me use the logo, which had been designed two years earlier by

Stacey Emerick. At the heart of my thinking was all the tax credit money we had been leaving on the table.

Tax credits are helpful only to people who pay taxes, and since Piedmont Biofuels, the incorporated co-op, had no tax liability, it could not avail itself of any of the state or federal tax credits that were becoming available for biodiesel. In a limited liability corporation, tax credits are allowed to flow through to the owners, which means that people with a tax liability, even if it is not related to biodiesel, can take advantage of the credits.

When news spread that I was about to take it up a notch, a miraculous thing happened: people everywhere started showing up with money. Some wanted in on the tax credits, some were simply enamored of biodiesel and some were under the misguided impression that every business venture I attempt turns out to be wildly successful. When I passed the million-dollar mark on commitments, I closed the investment entirely and to this day I turn down interested investors.

As all this was happening, I went to lunch with Dixon Harvey from Environmental Banc and Exchange of Baltimore. They are primarily a wetlands remediation company but they recently had made a significant investment in a plastics-to-energy company out west. He seemed to think that raising investment capital for alternative energy projects is very different from traditional investments, and that the hard part is not arranging the money but making the investment work.

That was an intriguing concept to me. Perhaps people have no problem investing in wind or plastic-to-energy or commercial biodiesel because there is a general unease in today's society that we may be approaching the end of the oil era. It could be that many people with investment dollars are delighted to jump into biodiesel — and that the hard part is indeed not raising the money but making the investment work.

There is a company in Golden, Colorado, called BBI. In the fall of 2004 they were a 27-person consulting firm led by a husband and wife team of owners. They are associated largely with Big Ethanol, but they currently publish *Biodiesel* magazine and have started poking around the edges of biodiesel.

My first brush with them was at a Department of Energy conference in Raleigh, where I shared the stage with one of their consultants. We were the two speakers who were to address biodiesel. He muttered into his coat and tie, drew some parallels between ethanol and biodiesel and by and large put everyone to sleep. I woke them up with my backyard story.

My second brush with BBI was when Carolina Biodiesel hired one of their ex-consultants to do a feasibility study and a business plan. The fellow came in from Colorado knowing virtually nothing about the local area and provided a nonsensical plan that looked like something off the shelf at Office Depot.

And my third brush with BBI was at the first-ever Virginia Biodiesel Conference, where one of their consultants gave a fantastic presentation on how to assess the financial feasibility of a biodiesel plant. He was great and his presentation was lively and solid, but I could not help seeing it as yesterday's thinking.

As a starting point, the entire presentation was tied to the price of diesel at the pump. BBI promised a ten-year study of the region, but a trailing ten-year study of the price of diesel does not necessarily say a thing about the ten years ahead. There appeared to be no peak oil calculations in their thinking.

Their graphs and data did nothing to differentiate biodiesel from petroleum diesel. They assigned it no green premium, no consumer acceptance, no sense that our air is polluted. They simply viewed biodiesel as a commodity, plugged it into their business plan modeling and declared, "Here is what you can and cannot do."

Fortunately for the crowd, the speaker immediately following BBI was Doug Faulkner, the founder of Virginia Biodiesel Refinery. He told the audience to forget feasibility studies, that oil is at $55 a barrel for a reason and that tomorrow's numbers will be based on $70-a-barrel crude.

Another shortcoming of the BBI worldview is that it is predicated on the old-fashioned belief that more volume is better for all. They showed a three-million-gallon biodiesel plant with a 5 percent return on investment and a 30-million-gallon plant with a 30 percent return. Big is always better. Except maybe not in biodiesel. Jeremy Rifkin may be on to something with his micronodal approach to energy production. Vivek claims that biodiesel has diseconomies of scale, which makes sense when factoring in the transportation costs. Waste vegetable oil often is contaminated with water, which means a behemoth biodiesel plant that is having to import waste veggie from near and far is paying to ship a lot of wastewater around.

The goal of Piedmont Biofuels Industrial is to make fuel for the local market from local feedstocks, just like a local winery. A million gallons a year is very small by NBB standards and today we are not sitting around with 30-million-gallon fantasies.

We may differ from almost every point in the BBI way of thinking. Their view is that biodiesel startups fail because they are too little, they are undercapitalized and they are poorly managed. I suppose we will need to stand by to see how our project flies before we will know who is right.

I also should say that the formation of Piedmont Biofuels Industrial cost me my seat on the board of Carolina Biodiesel. They are the folks in Orange County who called me up one day and described the abandoned salad dressing plant they owned. They were seeking a job creation strategy and, in keeping with our open-source philosophy, I freely told them everything I knew about biodiesel.

My own feeling is that there is no conflict of interest between our million-gallon plant in Pittsboro and their undertaking in Hillsboro, some fifty miles away. Every little town on this continent should have a million-gallon plant and every little town should strive to meet its own fueling needs. If we are ever going to be free of a top-down, government supported energy infrastructure that externalizes its true costs and operates in a monopolistic fashion, we need thousands of Piedmont Biofuels Industrials dotting the landscape of our economy.

fourteen

Sidestreams

A successful biodiesel reaction kicks out glycerin molecules that we don't have much use for. Glycerin is a crude, nasty substance that the biodiesel process creates in abundance and it is problematic as a sidestream.

At Piedmont Biofuels Industrial, we are proposing a million-gallon capacity, which will yield roughly 200,000 gallons of glycerin, or about two million pounds, a year. At a penny a pound for disposal to turn it into compost, getting rid of our glycerin stream will cost us about $20,000 a year, not counting labor and trucking.

When people hear "glycerin" they think "soap" and they are right about that. Procter and Gamble controls most of the worldwide glycerin supply, and they make, among other things, soap. They also have been known to thwart biodiesel legislation in Europe because we incidentally create glycerin that we have little use for. However, our glycerin is nowhere near pure enough for soap making. You could make soap out of it, but you would come out of the shower smelling like fries.

Commercial-grade glycerin not only is purer, it also often is subject to rabbinical food laws. Since our biodiesel is made from waste vegetable oil, we can't certify what it fried last, which means it never will pass the kosher test. And refining glycerin is an order of magnitude harder than making biodiesel.

Many people in commercial biodiesel apply glycerin sales to their economic models and some purveyors of biodiesel production technology like to top one another with claims that "our process makes a purer glycerin that you can sell."

My sense of this is "nonsense." If even a fraction of the biodiesel production that is proposed comes online, the crude glycerin market will completely collapse. Whenever I see biodiesel business plans that are based on glycerin sales, I discount them.

The reality is that glycerin is a by-product we need to deal with. It is pure carbon and can be a good ingredient in soil making. Some in the dirt business will take it for free, but most charge a penny a pound for disposal. At Piedmont Biofuels we have tried various strategies for glycerin composting, which I have written about in the blog:

Glycerin Composting

We have had a devil of a time composting our glycerin at Piedmont Biofuels. The stuff comes off as a liquid, 40 gallons at a time, and when we dump it in a designated spot it merely creates a giant Slip 'N Slide.

Successful composting apparently is an aerobic event, and when the pile is too wet the process is anaerobic, which is exceedingly slow. At one point Oneas and Farmer Doug from the College came to our aid and constructed vast piles of alfalfa, wood chips, horse manure and whatever else their recipe calls for. We set up a series of elevated bathtubs next to the piles and devised a system whereby worker members toted five-gallon buckets of glycerin from the reactor to the tubs, where it was stored for a time before going into the compost.

That was a big step forward. Oneas would take a colander, scoop a bucket of glycerin out of the bathtub (scooping is nasty) and run it through the colander onto some alfalfa to create a very popular goat feed.

Likewise, Doug would add the liquid to various "structures" prior to adding it to his piles. He is a composting machine, by the way, who regularly measures the internal temperature of his piles (keeps them at a constant 130 degrees Fahrenheit) and turns them three times a week to produce an end product that is "certified organic."

But despite all these efforts, we still are backed up on glycerin composting. The bathtubs don't have lids, which means it is easy for people not to cover them properly, which means the tubs fill up with rain, adding more liquid to our liquid.

One day when I was working in the new Executive Suite, Doug came in to have me take a look at this "weird substance" that had appeared atop his glycerin in the bathtub. I went out and examined it and figured it was biodiesel. We fetched Leif for a second opinion. Sure enough. Since we don't officially terminate the biodiesel reaction, it appears to continue making fuel in the bathtub.

We pumped off a five-gallon carboy and were delighted to have some "free fuel" coming out of the sideyard. This experience made me think we need to revisit our entire approach to composting glycerin. Leif and I agreed that an elevated conical, opaque vessel (similar to Alien Baby) would be ideal. We envisioned steps leading to the top, where we could top-fill the thing and let the fuel phase separate from the glycerin phase at its leisure.

"What we need are those conical units they are throwing away down at the chicken factory," said Leif. Having driven past them many times, I knew exactly what he meant. We jumped in the car and headed to the factory guardhouse.

No luck. I continued to pursue it, getting the manager's name and leaving persistent voicemails. When I finally connected with the guy, he turned out to be a neighbor of Cheryl's and had been following our progress through her. Cheryl was an early adopter of SVO, is a graduate of the biofuels program and has been a regular on the Energy Class circuit. She also has offered up her car as a demonstration model at repeated exhibits and can regularly be found pitching in around the refinery. It's nice to have a movement behind you when attempting to purloin some chicken factory cast-offs.

The huge conical vessels of our dreams were already spoken for and not to be had. But today I picked up two 250-gallon opaque totes with snug lids on built-in metal pallets. They are the way detergent is delivered in quantity to the plant, and the detergent supplier apparently never came to retrieve the "empties," which means we have added two beautiful totes to our composting operation. All we need to do now is transfer the glycerin from the tub to the tote and add more as it comes off the line and we should be able to repeatedly harvest the visible fuel layer off the top.

These units have fantastic ball valves at the bottom that should allow for easy draining. We could design a two-inch hose to attach to the bottom so that scooping will no longer be required.

If we are to cope with the impending glycerin glut, we are going to need new products based on it as a feedstock. Purdue University has done considerable work on using it as a de-icer for airplanes. Glycerin, after all, is an alcohol.

Backyarders struggle with glycerin. Some mix it with sawdust to make logs to burn, others burn it directly in Babington burners. My understanding is that it needs to be incinerated at exceedingly high temperatures or it gives off toxic fumes. I've heard of glycerin being used in brickmaking as well.

We took five gallons of our glycerin to Sam at Burlington Chemical to see if he could invent some sort of industrial lubricant from it. And we have taken samples to dirt makers and soap makers but to date no one is clamoring for more.

Since Piedmont Biofuels Industrial is entering a compound with extra empty buildings, we are hoping to invent some sort of product that can be produced onsite so that the waste output of the biodiesel plant can serve as a feedstock.

When the glycerin falls out of the biodiesel reaction, it takes methanol and potassium (our catalyst) with it. Reclaiming the methanol is fairly straightforward. When heated it vaporizes and when sent through a copper coil it is distilled back into a liquid to be used in the next batch of fuel. Getting the potassium out of the glycerin phase is another matter that we have not yet explored.

When fuel is finished we wash it with water, and the resulting wash water is another sidestream of the biodiesel process. We initially believed that our wash water would be rich in potassium but we were told the potassium was not "plant available" — i.e., would not come out of the wash water phase as a decent fertilizer. To solve this problem, we were told to run the wash water through a bed of vermiculite that would ionize the potassium and make it a splendid fertilizer.

Most of the vermiculite in America is imported from China. It is a flaky mineral substance. I managed to find a mine in South Carolina and bought fifty pounds of the stuff, but the co-op members who initially expressed interest in exploring our wash-water sidestream have been swept away on other projects. Suffice it to say that wash water is on our list of projects and, as with the glycerin, we would like the waste to be the feedstock of another product. We are working on it.

fifteen

The Regulatory Layer

O ur state has trouble with the fact that people can make their own fuel, and there typically is no differentiation made between vegetable oil and rocket fuel. If it powers a vehicle, it is regulated as fuel. When I tell the local authorities, "You can't light it with a match," they don't care. Fuel is fuel.

When I say, "It's nontoxic. It can be composted. It's not a hazardous material. If you drank some you wouldn't die," they don't care. Chances are very good that biodiesel will stay in the "fuel" section of their book.

When I first went to Chatham County government with my plans to start making biodiesel in the sideyard of Moncure Chessworks, they flipped. They not only forbade fuel making, they also deemed that my sculpture business, which had been there for six years without a single incident or complaint, would have to shut down.

With that threat looming, I went to work on getting a conditional use permit for Chessworks, which turned out to be a long and arduous journey. The county ended up changing the zoning ordinance to allow "art fabrication" in areas zoned for business, which was an accidental victory for all our visual artists.

But it was clear that biodiesel production would have to go on somewhere else. "Go to your house, which is not zoned, and don't tell us about it" was the advice of one county official. Which is exactly what we did. Our current operation is still in an unzoned portion of the county, keeping us in compliance with the planning people.

We are in an abandoned trailer, which apparently is a violation of the North Carolina Commercial Building Code. When I asked why we could not conduct business out of a trailer, I was told that the structural loading requirements for a business are different from those for a residence. When I pointed out that our operations are on a cement slab on the ground and that I didn't believe there were any structural issues, I was told I would have to get that certified by a professional engineer.

My old architect friend has come to my rescue on a number of occasions over the years, helping me get certificates of occupancy at the last minute and guiding me through the maze of regulations that surround buildings in general. One day I caught him on his Harley Davidson at the post office in Pittsboro. Instead of wasting his time providing his professional stamp on a cement pad, he remarked, "Farm buildings are not subject to the North Carolina Commercial Building Code."

And with that I was off to the USDA. Rachel and Oneas already had started the process of registering as a research farm for the feedstock crop work they were doing so I accelerated that process and got us a farm serial number.

As soon as articles started appearing in local papers about our fuel and our operations, county government came calling. Planning noted that we had done what we were told, but inspections went after us because of our trailer. When we played the farm card, we were covered and the conversation fell silent. I never had the chance to say, "Roger, get back to me when you regulate agricultural co-ops," but I was very glad to know we could summon the forces of agriculture to our aid if need be.

So our operation is legal with the local authorities, which is a tremendous blessing. There are many backyarders who operate in fear of the local authorities because of zoning issues. People go off-grid to protect their operations from having to get power permits. People go to great lengths to hide. We have not hidden and we have kept it legal. Yet from a regulatory standpoint, the local authorities are but a small part of the swamp.

The much larger issue is staying legal while running on homemade fuel.

The main hurdle is the EPA. They regulate fuels in the United States, and they do not care if people make small quantities of fuel and put that fuel into their own tanks and drive down the road. From an environmental standpoint, that is allowed. A problem arises when you want to sell your fuel for onroad use. That simply is not legal without EPA registration, NBB membership and verification that your fuel meets the biodiesel specification.

The USDA regulates weights and measures. Although most backyard biodiesel is distributed in approximate quantities ("That looks like five gallons"), it is illegal to dispense fuel to the public without an expensive USDA-compliant meter.

Taxes are another story. The IRS wants its cut, in the form of an excise tax, of anything that goes into a fuel tank. To register to pay the tax, a Form F720 is required, and to use an F720 you require an EIN number, which is easily obtained online.

The IRS has visited several members of our co-op who were running on straight vegetable oil. One did a prominent television spot about his conversion and another put up a website demonstrating how easy conversion is to do. A diligent taxman from South Carolina visited everyone mentioned on the site, and for what it is worth apparently he really did represent a "kinder, gentler IRS." He nicely showed everyone how to fill out the forms, collected the back taxes and offered guidance to Rachel and Leif on covering the IRS position in their biofuels classes.

In Canada, where government appears to be more enlightened, federal excise tax on biodiesel has been waived, as has the provincial tax in Ontario. Illinois also has granted biodiesel a holiday from state taxes. We have been working hard on getting North Carolina to do the same.

Staying legal looks pretty simple: set up in an unzoned part of town, register as a farm, don't sell homemade fuel and voilà — you are legal. But selling the fuel could generate the revenue stream necessary to support the local fuel maker, which would allow biodiesel to start contributing to local economies.

My frustrations with finding my way through the regulatory layer prior to the previous upscaling of Piedmont Biofuels are well-documented in the blog, which includes discussion of an inspired speech by Kim Kristoff:

Regulatory Swamp

It appears that Piedmont Biofuels will soon be moving out of Summer Shop. Summer Shop is a sentimental and magical place for me. It was recovered from poison ivy and garbage. At its heart is an old pole building that has been wired and rewired and wired again. It's the place where I once strapped a new set of oxygen and acetylene tanks to a pine tree to begin my sculpture business. And it is where much of the work for Piedmont Biofuels began.

The absence of walls and good electricity made me leave Summer Shop and open Moncure Chessworks, five miles away in the village of Moncure. And for similar reasons, plus the cold, the long, twisted lane that is not good for tanker trucks and the sheer volume of traffic about the place, it is time for Piedmont Biofuels to move on.

It's both an exciting and a mournful time. I like the proximity. I even like the traffic. But it's time for me to think of Summer Shop as an incubator, and to realize that Piedmont Biofuels is about to hatch, just as the sculpture business once did.

After speaking at the NC Biomass Conference, I received an e-mail from someone I had never met. Attached was a clipping on a guy in Atlanta who started making biodiesel and supposedly was fined by the EPA to the tune of $25,000 a day.

At this point, that's more money than we have invested in Piedmont Biofuels. Since the move is about to change all that, and since we are about to start spending in earnest on this project, I thought it best to investigate the regulatory side of the business.

I started by sending an e-mail to the EPA. The EPA kindly sent me off to the North Carolina Department of Environment, Health and Natural Resources (DEHNR). I poked around DEHNR's website, and after finding nothing on biodiesel I entered the "One Stop Permitting" department. I made a call. The poor woman had never heard of biodiesel and was afraid to say anything about anything. She said she would investigate and call me back. Never did.

Then I went to the glorious opening of the Alternative Fuels Garage. Alex Hobbs and a panel of dignitaries spoke, and the keynote speaker was Kim Kristoff, CEO of GEMTEK and founder of the Bio Products Manufacturers Association.

Kristoff's talk was both irreverent and enlivening. He was a complete inspiration. At the heart of his talk was the idea that anything that can be done with fossil fuels can be done with biological products. His company has made solvents, adhesives, even car doors from nontoxic sources.

He basically got up onstage with the blue ribbon panel from NC State and threw away the text. "Your government does not share your agenda," he said. "Your government is lying to you."

Listening to him I regretted the endless hours I had spent researching the proper "permits" required to upscale Piedmont Biofuels. When he was done, presumably he was whisked away to his private jet, but I was able to buttonhole the head of NC DEHNR, with whom Kristoff had shared the stage.

I explained to him what we were doing, how we were about to scale up and how his agency had been little to no help, and I asked point-blank, "Should I proceed with trying to stay legal or should I heed Mr. Kristoff's advice and just do it?"

"I think it would be best to continue working with us," he said. We exchanged cards and left it at that.

I came home from the ribbon cutting at the Alternative Fuels Garage on fire to effect societal change. I sent an e-mail to Kim Kristoff applauding him for his amazing performance. I redoubled my efforts to get answers out of NC DEHNR. I went over to the future home of Piedmont Biofuels and chain sawed a downed hickory that was blocking our entrance. I imagined a campus that included a refinery, a truck and trailer rental area and a biodiesel bed and breakfast that could be a model for ecotourism. I saw us heating the entire place with solar thermal energy. I saw us developing a model of sustainability. New construction would be straw bale. New wiring would be 12-volt for photovoltaic systems. Fuel makers would walk to work. It would be perfect.

And then I went back into the regulatory swamp. In order to stay clear of the EPA we would need to join the National Biodiesel Board. A small-producer membership would cost $2,500 dollars, a quarter of our current budget. The sheer corruption of the process hit me hard. I sent an e-mail to my brother Glen, telling him I was simply going to reopen in São Paulo, where the corruption is clearer, rather than join the National Biodiesel Board. I was discouraged that night.

Glen wrote me back an amazing e-mail that buoyed my spirits. He talked about his discouragement with his wind turbine project, about unfulfilled promises made by governments along the way. And he made a startling observation: "You might as well play golf with fat white guys. It would be more fun than playing with the nipple-ring crowd. At least fat white guys like playing golf."

He went on to say that because yellow grease would always be local, there would always be a role for local producers, and that the guys from the National Biodiesel

Board should hear the voices of the small producers. In response to my query as to whether or not we should join he said, "Go for it."

I woke up this morning all ready to join the National Biodiesel Board. And this afternoon Rachel and I went off to Triangle Clean Cities to hear one of the NBB's people speak.

After hearing two hours of singing the praises of biodiesel, its reduced emissions, its increased lubricity, its being a source of independence from foreign energy sources and all, I asked, "Since the public cannot buy biodiesel in North Carolina, your industry has spawned a whole bunch of underground backyard home-brewers like me who just want to run on biodiesel and are forced to make their own. Now that the EPA is cracking down on backyard operations like mine, can I buy a membership in the National Biodiesel Board and get some cover from the EPA?"

Not really. The $2,500 small-producer membership entitles us to the data from the human health effects testing they have done on biodiesel as a fuel. Not our biodiesel. Theirs. It's a joke. I drove to the meeting on straight vegetable oil. That's also illegal? What am I supposed to do, drop two million on health testing of this alternative fuel? How about this? "Officer, you can cook with it."

On the way home I could not help thinking of Kim Kristoff. In one of his e-mails to me he said, "Keep on keepin' on .... Let them tear the ears off your favorite dog before you succumb to the tortures of biodiesel registration and unfair taxation ... but do expect threats and some grinding of teeth. We are talking about the most debt-ridden institution in recorded history ... at least since the plague years ... the US government ... and this government is blind as a rock to biodiesel."

It's confusing. I'm not sure what to do. There once was a time when slave owners were not allowed to let their slaves go. Is this such a time? Didn't Martin Luther King teach us that it is immoral to submit to an unjust law?

I'm wondering if Piedmont Biofuels needs to go forward with its move. We have $10,000 on hand to pull it off. Perhaps we should not waste a nickel in the regulatory swamp and instead invest the resources in making a cleaner-burning alternative fuel from waste vegetable oil.

That entry may have been a little dramatic. I've learned a lot since then. A thick regulatory layer sits atop biodiesel. It is comprised of local regulations, state and federal tax regulations, ASTM fuel standards and EPA human health effects standards. Though this layer can be daunting to the newcomer, it can be successfully managed. At Piedmont Biofuels we have spent considerable effort at staying legal so that we can talk about what we are doing without fear of running afoul of the authorities. We advocate this approach and we are proof that it works.

sixteen

Quality Control and Analytics

A tremendous hurdle facing the backyard brewer and the small producer alike is making fuel that conforms to ASTM specification D6751. This is not a specification for B100, or what is called "neat" fuel, but for biodiesel to be blended with petroleum diesel.

Achieving the specification is not hard. We can make ASTM fuel out of a dead squirrel on the workbench. The hard part is analyzing the fuel once it has been produced. The equipment necessary to perform all the assays costs more than most backyard and small-producer operations are worth.

There are fifteen tests that need to be performed, from flash point to water and sediment content to cloud point to phosphorus content. The numbers bandied about by Iowa State and the commercial biodiesel makers for the cost of the gear required to run this battery of tests are in the $75,000 range. Kumar, at Yokayo Biofuels in Northern California, has been collecting used analytical tools for quite some time and guesses that if he waits long enough, goes to enough online auctions and keeps his ear to the ground he should be able to assemble everything he needs for a large-scale biodiesel production company for around $25,000.

Kumar is a quality fanatic. Most National Biodiesel Board members do not test every load of outgoing fuel. Most certify a process, whether batched or continuous flow, and do ASTM certification on a sample to ensure they have it right. Those without the necessary investment in analytics, or those unwilling to invest in the time to do proper testing, can sample as few as four times a year.

The law requires biodiesel to conform to the ASTM specification, but since there is no one, not even the National Biodiesel Board, to monitor or police quality it is not unusual to see whole shipments of "goo" go from manufacturers to distributors and end up in the fuel tank.

Unfortunately it often is the grassroots organizations that have to deal with the problem of "off-spec" fuel since the end-user generally shows up at our door. Surely the most famous discussion of "goo" occurred in the spring of 2004 at the National Biodiesel Board's inaugural conference in Palm Springs, California. Girl Mark showed up with a little plastic vial of a reddish, syrupy-looking substance that had slipped through the California B100 community under the name of "biodiesel." She carried it with her and pulled it out of her sleeve whenever the occasion demanded.

I believe the suspect product came from a plant in southern California. It may have been the same plant that offered tours at the conference. Rumors circulated that girl Mark was threatened by legal action for her public declarations of poor quality and her constant willingness to produce her vial of "goo."

There was a famous confrontation in the lobby of the Wyndham Hotel that I covered in Energy Blog:

The Joe Jobe Experience

As I walked through the lobby I found Leif and another guy having a quiet conversation with Joe Jobe, executive director of the National Biodiesel Board, and one of his California consultants. I crashed the conversation and away we went.

We talked about open-source philosophies, and I felt Joe Jobe was listening. At this point girl Mark entered the conversation, taking it up a notch. She tossed out a vial of bad fuel made by an NBB member and waded into the quality issue. She clearly is interested in driving a stake through the heart of the argument that small producers "are mixing up just anything."

Something you have to understand about girl Mark. She's intense. Once she bites down, she does not let go. She knows more about biodiesel than most of the 600 delegates at the conference. And more about diesel engines. And more about the industry. Best strategy I can think of for dealing with girl Mark? Don't mess with her. She will chew you up and spit you out.

Poor Joe Jobe was conciliatory and candid, and he pushed back as best he could. I tried to throw in some affirmations. And the odd punch. A crowd gathered around the table. It was like a schoolyard fight. Big Soy stood on Joe Jobe's side and home-brewers and small producers gathered on our side. Jake, a researcher from Texas, continually butted in with insights about how wonderful the mere dialogue was — and he was right.

Girl Mark was armed. She had arbitration, mediation and litigation on her side, and she punctuated it all with the vial of "goo" which sat in the middle of the table. She was a fierce warrior in the heat of battle.

And Joe Jobe persevered. He gets a point for withstanding the onslaught. He strikes me as genuine and as a true believer in biodiesel, someone who is honest and fair and just wants to have a good conference to further a good industry. He appears to have a heartfelt attachment to farmers.

And girl Mark listened too. She strikes me as someone who craves facts and the truth and as a true believer in biodiesel. She seems honest and informed, someone who just wants a fair shake for the small producer. She might tell you she was swimming in deep water full of sharks, but Joe Jobe wasn't exactly lounging on the beach. By the time the discussion ended there were forty or more people around the table, and there was a round of applause. Those forming the circle were not chanting, "Fight, fight, fight" as they did in elementary school. Rather they were chanting "Agree, agree, agree."

The problem for the backyarder and the small producer is that they cannot pound a sign in the yard that says "Biodiesel for Sale" unless they can demonstrate that their process generates ASTM-quality fuel. Sending a sample off to a commercial testing lab costs anywhere from $800 to $1,000, again out of reach for most of them.

I had a terrific conversation about this one day with a fellow from Asheville and I published it in the blog with his permission:

The Analytics Piece

Yesterday I was working on our Clean Technology Demonstration Project (our friend Grant) and a call came in from Matt at Blue Ridge Biofuels in Asheville. He wants to know about our tank truck and about taxes, and I pace around the front yard as I talk. It's a typical "open-source" phone call in which I divulge everything I know in an effort to help the movement along.

I've never met Matt, but he's informed and articulate and we got onto a comfortable beam about all things biodiesel. Among other things, he tells me that Blue Ridge Biofuels was named Mountain Biofuels for exactly one week before a hurricane passed through Asheville.

That interests me because my six-year-old son Arlo popularized the name one night when he was drawing a picture on the kitchen floor. At the table were Chris Jude from Boone, Mac from Asheville and others. We were talking co-op structure and reactor design and the usual obsessive stuff, and what Arlo took in was that people were going to be opening biodiesel plants in the mountains. So he drew some mountains, most with snow-capped peaks, and added bicyclists and rock climbers and refineries. Normally I slip such masterpieces into the woodstove when I clean the kitchen, but I set his drawing of "Mountain Biofuels" aside and later handed it to girl Mark, who was on her way to Asheville to do a workshop.

Soon after, I was delighted to see an e-mail fly by from Solon, polling their co-op members on the virtue of changing their name from Asheville Biodiesel Cooperative to Mountain Biofuels, and apparently the name stuck. For one week. A hurricane caused flooding in Asheville, and a petroleum outfit named Mountain Energy had a horrific spill into the local rivers. Not wanting to be associated with such poisoning, the folks at Mountain Biofuels became Blue Ridge Biofuels.

Arlo missed all this. His brief moment as an "adman" will be forgotten, which is OK since he is quite prolific with both art and ideas.

Matt and I covered the usual terrain and arrived at my current fixation: biodiesel analytics. I had heard that Blue Ridge Biofuels had the analytical part all figured out, using lab equipment at Warren Wilson College, and I was exceedingly stoked. That turned out to be untrue. In fact, all they have is some guy who is a chemist "who knows how to do all that stuff," which is about what every backyard operation everywhere has.

Rachel and I currently are pushing Central Carolina Community College to add biodiesel analytics to their bioprocessing program. We want them to get all the expensive gear and teach folks how to run the assays necessary to determine ASTM specification on biodiesel. We want that to become a free service for backyarders across the state.

My point to Matt was that in almost every backyard operation I am aware of there is someone who would become a professional fuel maker if only they could sell their fuel. And if every backyard operation turned pro, there would be some commercial production in this state.

He argued that getting ASTM-tested is not that hard. Send your sample off with $800 and voilà. Get it tested four times a year and spread the cost over 50,000 gallons and it's no big deal.

Girl Mark's concept is that you could assemble a backyard operation that would produce 50,000 gallons "in buckets." Piedmont Biofuels could easily do that with our current operation. All we need is the labor, and to afford the labor we need to be able to sell our fuel.

The problem with Matt's "test it four times a year" is that it leaves the backyard producer in the same boat as the current commercial producers, who happily ship off-spec fuel with impunity. At Piedmont Biofuels we currently have 110 gallons of "goo" that we can't even re-react that came from a commercial producer in Tampa.

What we need is a college or some other institution to step up and make testing available. I almost got some traction from the National Institute of Environmental Health Sciences in Research Triangle Park to help us out on the analytics front, but in the end they didn't want to get the column on their gas chromatograph oily by putting biodiesel in it.

It was a great flap with Matt from Blue Ridge, and I hope I put some pieces in place for the analytical part of the biodiesel puzzle.

The reality is that proper testing doesn't happen at the commercial level either. I have a theory that an aspect of quality control that is missing at the commercial level is simple visual inspection. In the backyard everything is done by visual inspection. The recipe is set by the definition of the line between the fuel and glycerin phases. Glycerin typically is drained off the bottom of a conical tank, and the draining stops when the dark glycerin ends and the light biodiesel appears. Wash water falls to the bottom of the tank and typically is white, or at least lighter than the amber fuel.

The backyarder likely is unable to test for free and total glycerol using a gas chromatograph but possibly could get a sense of the clarity of the fuel by holding a mason jar up to a window and trying to read a newspaper through it.

Industrial production, on the other hand, takes place in stainless steel reactors and black pipes and the "visuals" are provided by chemical plant instrumentation, which means a "bad batch" can pass unnoticed into the world.

The most notable shipment of "goo" that came our way arrived with a Clif Bar cross-country marketing tour. Rachel had been getting some calls. Apparently a brand new diesel truck had rolled off some production line in the Midwest and been snapped up by the makers of Clif Bars. It was beautifully painted with company logos and slogans like "Zero Net Emissions" and "Running on 100% Biodiesel," staffed with a couple of keen young marketing executives and sent out on tour to give away Clif Bars at festivals and celebrations around the country.

Clif Bars are little, dense, compressed affairs of organic material that are supposed to give consumers energy and lift and keep them from cramping up during prolonged periods of physical activity — long bike rides, climbing tall rock faces and other extreme sports. Apparently the Clif Bar market niche is that they are organically produced.

The good people of Clif Bar set out on their B100 journey with a pair of 55-gallon drums strapped to the back of their truck so they could stay on B100 through the dry spots. Their problem was that they filled up at a National Biodiesel Board member in Tampa they had found on the NBB website, and they ended up on the side of the road in a one-horse town in Florida.

The mechanic they turned to took a couple of weeks replacing their entire fuel system (had to order the parts), and the whole affair was blamed on the fact that they were running on B100. They had spoken to their engine manufacturer and been told that running B100 voided their warranty.

By the time they arrived in Pittsboro, they were hot and bothered and strung out on biodiesel. They were running on petroleum at the time, and the moral incongruity of driving around on petroleum in a vehicle that was painted up with the virtues of biodiesel was too much for them. Rachel had to teach, and directed me to draw a fuel sample from their storage tanks.

I siphoned off some fuel in the parking lot, held it up to the sun and could tell right away that what they had been running on was not biodiesel. This was a simple visual inspection by a backyarder. No wet chemistry. No near infrared. No gas chromatography. The sample was just "goo," identified by someone who has made his share of the same.

I invited them into the air-conditioned office and asked them a few questions. They knew nothing about diesel technology, nothing about biodiesel. They simply wanted to drive around the country to promote their little energy bars, and they wanted to be congruent along the way.

We spent almost an hour on the white board, and they drove out to the refinery to unload their "goo" and to fill up with store-bought B100 from us — fuel we had bought from World Energy, which also has a plant in Florida. For over six months we have been hoping to re-react the 110 gallons of Clif Bar "goo," but all attempts have failed. My guess is that it will go to the compost the next time we are critically short of 55-gallon drums.

I sometimes outline fuel quality on a spectrum. The worst possible fuel for a diesel engine is straight vegetable oil. It will lacquer combustion chambers and coke fuel injectors over time. Next worst is unwashed homemade biodiesel. Everyone who makes their own has run on it. Josh Tickell didn't even advocate washing in his early work. Many people view washing as optional — including me when the fuel gauge is on empty. Next on my imaginary spectrum is fuel that has been washed. Next to that is fuel that has passed ASTM testing. But, as Todd Swearingen has said online, fuel quality really is dictated by the fuel maker. Not the lab, not the equipment, but the person doing the work.

Whether it's "goo" in a vial presented by girl Mark or 110 gallons in the back of our shed or the crap Kumar has kept from entering his customers' tanks by doing analytics on the hood, the fact is that commercially-made biodiesel does not have a highroad claim on quality.

At their 2004 annual meeting, the National Biodiesel Board passed a resolution that allowed for a new membership category called "small producer." As the vote passed, a fellow from Archer Daniels Midland rose and spoke against small producers everywhere. His concern was that small producers could not spend enough on safety or quality and should not be allowed access to all the benefits of the National Biodiesel Board.

The resolution passed anyway, and that night a dozen or so small producers celebrated around several tables together in the hotel bar. Two remarkable things happened that night. The first was that a complete round of drinks arrived unannounced, sent to our table by an NBB member who had produced the famous vial of goo. He gave girl Mark a nod and a wink as she looked around to see who had sent our grassroots crowd free drinks.

The other remarkable thing was that my daughter Kaitlin joined us. She was twelve years old at the time and was visiting from her home in Iowa. She had been a "Celebrity Guest Author" one night in Energy Blog, and every single person at the table had read her now-famous "Letter to the President." She sat on my lap and relished her ten minutes of fame.

Letter to the President

Dear Mr. President Bush,

Hi, my name is Kaitlin Estill. I live in Ames Iowa and Moncure North Carolina. At my Dad/Mom's house. Both are small insignificant places compared to Washington D.C. I'm 12 years old. I have a stepmother and a father, a stepfather and a mother, so I am fortunate to have a large family. With that large family come many political views and so forth about the economy, environment, the war in Iraq etc. So please do not think this letter is coming right out of my parents' views. I have formed my own opinions about things too.

First of all I'd like to say that no one is perfect and I can't blame one single person on this earth for all his or her mistakes. Everyone makes mistakes. We wouldn't be human if we didn't; we also wouldn't learn. Second of all I want to say that I realize your job is probably one of the most stressful in America.

Still ... I am literally embarrassed to be a part of this country right now. This is a feeling I have never felt before in my 12 almost 13 years of life. I almost thought about living in Canada, which has its issues too. I'm not trying to pretend that Canada is perfect but right now it seems a lot better than here. You may ask, "Why pick Canada?" Well, actually there are two reasons. I have extended family in Canada that I would love to see more often. The second reason would have to be their superior candy selection to the United States.

My favorite candy in the world, Aero bars, can only be found fresh and new in Canada. Although I could talk about candy all night and well into tomorrow there are more relevant issues at hand.

For instance, do you care about our planet? About our endangered species, some of which (most likely dolphins in this case) are probably smarter than you?

As I mentioned before I have lots of family so we go on many family vacations including Bahamas, Mexico, Key West, and numerous other places, so I have gotten to see some wonderful wildlife. As I have seen beautiful wildlife my expectations have gotten higher and higher, so when I see some animals or plants I still accept their beauty but I want to see more than just that. I want to see more and learn more.

I am very lucky because I get more opportunities than some people do. I have seen a dolphin before — actually about five or six — but I haven't ever seen a sea turtle. I fear I may never get to. I know that one 12-year-old's wishes are not going to make an impact on you, but pause for a moment and think about this: what about all the 12-year-olds in the world not getting to see a sea turtle and many other animals on the endangered species list?

If you think about that then why not think about all the people in the world that are not getting to see and live and enjoy this exquisite planet? A single factory is giving off so much pollution, and you are giving permission for pollution. Forget about the money part and realize that stopping this kind of pollution isn't just for a few species. It's for our entire planet!

I could say, "Heck, why am I complaining?" I could live a century from now and pollution and using all our natural resources would leave me with not much and not getting to enjoy our wonderful planet half as much as I do now.

My Dad makes biodiesel, and runs his truck on straight vegetable oil. It is becoming more popular now and people are catching on to it. This is one of the few good things my Dad does for the environment. Please forgive me if those words aren't in your vocabulary.

Biodiesel is also a good way to make sure we are not using all our natural resources so in years to come we will have conserved them. That sort of brings up the Iraq war issue.

OK, let's just say I quit the politeness for a second and act like a normal 12-year-old: dude, what's up with Operation Iraqi Freedom?

Did we find any weapons? I don't think so. Also some people could argue the case that since America has these weapons can we really trust someone who honestly should be in jail right now for numerous war crimes, is ruining our planet, completely screwing up education and healthcare and has such deadly weapons?

So basically at the end of the day we ask ourselves why are we still in Iraq? When it really comes down to it, we want the oil, which I think is unnecessary and we should leave the poor people alone.

They hate us. We have to admit that we're not helping them — unless your definition of helping them is having tons of people killed every week and if that is the case please let me know. Yes, getting rid of Saddam Hussein was definitely a plus. I will give you credit for that.

So can we both admit that America's greed for oil is keeping us in Iraq? I think the first step is being truthful with yourself (this includes all issues in America) and then being truthful with our country, America.

Sincerely,

Kaitlin Estill, Concerned Citizen

While there is little doubt that the group of back-yarders and small producers gathered around the table in the DC hotel bar was delighted that the NBB had voted in a new "small producer" category, the fact remains that quality control in the biodiesel industry still rests with the consumer. Since there is no policing or watchdog effort going on, large and small producers alike are free to ship whatever they want and call it biodiesel.

The NBB does have a pilot quality program called BQ9000, which sets rigorous standards for record keeping, manufacturing, storage and use, but it is in its infancy and even when fully rolled out it will not put the NBB or anyone else in a quality-monitoring role.

There is a Canadian equivalent standard called Biodiesel Driven, which generally is considered to be even better than BQ9000, but few producers have adopted it at this point.

For consumers, fuel quality seldom enters the conversation unless, as with the Clif Bar marketing tour, bad fuel leaves them on the side of the road. There is an expectation that biodiesel is simply biodiesel, and the real concern in the mind of the consumer is: "What will it do to my engine warranty?"

In the United States, the Magnuson-Moss Warranty Act protects consumers and engine makers alike by clearly stating that a manufacturer can warrant only parts and workmanship. It cannot, therefore, stipulate or object to what you use for fuel.

But lend your brand-new diesel Mercedes to your friend and have him fill it up with unleaded and destroy your fuel system and the dealer is unlikely to honor any sort of warranty. If you get water in your unleaded fuel and put it into your brand new engine, the resulting damage will not be covered in the "bumper-to-bumper" warranty either.

The Magnuson-Moss Warranty Act is small comfort to the B100 enthusiast who has to invoke federal law in order to get repairs done at a local dealer. In reality, of course, most local dealers have never heard of biodiesel and are unable to ascertain what fuel the vehicle has been running on.

Quality control and analytics are the biggest issue facing biodiesel in America today. Poor quality affects the reputation of the fuel, and when off-spec fuel shows up everyone from the backyarder to the commercial producer is hurt. To ensure quality fuel we need the collaboration of academics, the National Biodiesel Board and those in the backyard. The consumer must have only one experience with biodiesel, and that is flawless performance.

seventeen

Biodiesel Economics

When we started making biodiesel in the backyard in 2001, we figured our costs to be about $0.50 a gallon. In those days methanol was procured by driving a pickup truck to the dock of the chemical distributor and tow- strapping a pair of 55-gallon drums in between the wheel wells. (We later came to learn that carrying more than 100 gallons of methanol at a time is completely illegal in North Carolina.) Our rough calculations indicated that if we added a labor component we could make biodiesel for around $1 a gallon.

Our cost for methanol has increased by four times since then, although that now includes delivery. And we've moved from endless little plastic bottles of Red Devil lye to potassium delivered in bulk. While prices have edged up, the general backyard consensus pegs the cost of homemade at somewhere between $1 and $1.50 a gallon, including labor.

At the time, store-bought B100 was selling for between $3 and $3.50 a gallon and petroleum diesel was around $1.69. Why commercial biodiesel was twice the price of petroleum diesel was one of the great mysteries of the world to me.

My first clue to solving this mystery came from Gene Gebolys, president of World Energy, at the Fuel Choice Conference in Raleigh. At the time, World Energy was the largest distributor of biodiesel in the US. An excellent speaker, in the courtyard of the MacKinnon Center he explained that the price of biodiesel was at an all-time high because Congress had not yet passed a farm bill.

What? Apparently the farm bill authorizes funding to a USDA subsidy program run by the Commodity Credit Corporation (CCC). It's a program designed to subsidize producers of bioenergy for a portion of their feedstock costs. Both ethanol and biodiesel producers can jump into this trough. Right below the surface of the CCC lies a hairy world of agricultural subsidy that is vastly complicated. Subsidies vary based on the market price of soybean oil. Subsidies for waste vegetable oil and animal tallow feedstocks are about 30 percent of the virgin soy amounts and are pegged to the price difference between virgin soy and waste oil. The subsidy is based on incremental production over the previous year.

Let's say you have a plant that is capable of making two million gallons of biodiesel per year. And let's say the CCC will pay for half the cost of your incremental feedstock. If you made a million gallons last year, your increment for this year could be based on an extra million gallons. But as you are not going to get paid on that increment until the farm bill funds the CCC, it is against your best interest to make one drop of fuel more than a million gallons. Basically, the commercial producers, who are all lined up at the CCC trough, stopped making fuel while they waited for Congress to pass the farm bill.

That explanation of why biodiesel was priced so high was almost too wacky to believe, but when I studied biodiesel production at Iowa State, sure enough, they had plant designs based on slowly ramping up incremental production in order to maximize CCC subsidy.

In 2004, this program was a line item of over $100 million in the federal budget. In 2005, the Bush administration submitted a budget that would limit outlays to $40 million. Either way, it is a lot of soft money. I call it that because it is not funded in perpetuity and could run out any time, making its existence difficult to factor into the price of biodiesel. When discussing the price per gallon should we quote subsidized feedstocks or not? And when deciding whether we should enter the business, should we pencil in a soft money kickback for our feedstock costs?

The price of biodiesel does not seem to follow market forces. Soy oil is traded on a market, so one would think the price of biodiesel could rise and fall with the price of virgin soy. Not always. Waste vegetable oil is traded on an open market, so one would think the price of biodiesel might follow that. Wrong again. The price of petroleum diesel goes up and down, presumably based on market conditions, but the price of biodiesel seems immune to shifts in the petroleum market too.

The glaring omission in the backyarder's calculation is that feedstock costs money for the commercial producer. Used fryer oil generally is abundant and is free to the backyarder but it generally is sold by the pound to the commercial producer. Feedstock can range from $0.07 to $0.15 a pound for waste vegetable oil and all the way up to $0.30 a pound for virgin soy. If we pencil in eight pounds per gallon (assuming there is some water content), feedstock costs come in anywhere from $0.56 to $2.40 a gallon, to which we need to add methanol and potassium or lye.

Assuming for a moment that we stay in waste vegetable oil, and assuming that we can get a million gallons for $0.10 a pound, suddenly the $1.50 a gallon calculation that worked for the backyarder shoots up to $2.40 a gallon for the commercial producer. And none of this includes taxes. In North Carolina, both petroleum and biodiesel carry about a $0.50 a gallon tax, so if we can make a product for $2.40 and add another $0.50 to stay legal, we are coming out of the chute at around $3 a gallon. That's cost. If we want to make a profit selling our fuel, let's sell it for $3.50 a gallon. That means that every time we make and sell 1,000 gallons we pocket 500 bucks.

And $0.50 a gallon is an incredibly high margin in the fuel business. I remember listening to Kimber from Biofuel Station, one of the first B100 distributorships in Northern California, lament the fact that she was living on a $0.06 a gallon margin. Almost two years after the amazing story of how she got the local authorities to allow her to open her business and stay legal, I entered into negotiations with the Jordan Dam Mini Mart in Moncure. A significant operator on the edge of a busy four-lane highway, he is lucky to get $0.06 a gallon on his petroleum products. Apparently the profits in the gas station business are in the chips and cigarettes that sell along with the fuel.

When we started selling street-legal fuel in the spring of 2004, petroleum diesel was around $1.69 a gallon and our price was $3.50 a gallon. In the summer of 2004, petroleum diesel hit $2.25 and we remained at $3.50.

$3.50 a gallon is not expensive. Petroleum would be much higher if it weren't so heavily subsidized. By now, I have almost grown weary of my own "externalities" speech, which I have given dozens of times since we started selling the fuel.

Many people are drawn to biodiesel for its backyard economics and are shocked to learn that unless they start making it themselves they are not going to save money by using the fuel. Some understand how petroleum has externalized its true costs and others do not.

Those who come to biodiesel to relieve pressure on their wallets generally go away at this point. Most others stay on and buy fuel. We saw a dramatic increase in co-op membership and fuel sales after the federal election of 2004. People woke up on Wednesday morning, stared ahead at another four years of the Bush-Cheney-Big Oil administration and started mapping strategies to get off petroleum.

In our narrow and intense B100 niche, these externalities are well understood and people happily pay a premium to be free of petroleum. In the blog I have addressed the pricing issue over and over, ranging from utter indignation to sympathy and back again. Here is a telling sample:

Wood, Fuel, Eggs, Garlic

I had a chance to go out to the message boards and relevant blogs tonight, and there was an entry on Kumar's *Fueled for Thought* that caught my interest.

He had a reader post:

CONSIDERING BIODIESEL CAN BE MADE FOR ABOUT $0.54 WITH WASTE VEGGIE OIL, HOW CAN YOU JUSTIFY THE PRICE AT THE FUELING STATION? TO REALLY GET PEOPLE INTO THE ECO-SCENE THE COSTS MUST COME DOWN. THEY CAN IF EVERYONE IS NOT LOOKING TO MAKE AN EXCESS BUCK.

I QUIT "HOMEPOWER" WHEN THEY WENT SLICK.
WHAT COLOR IS YOUR VOLVO? D.

This struck a chord in Kumar. He explains about capital costs and labor costs and the economic reality of biodiesel. Although he is a lever for societal change, he is paid like a grocery clerk. And he is sick of the $0.54 myth.

I read this exchange and thought about heating with wood. I live in a woodlot that casts off more trees every season than I can harvest. I simply find them down in the woods and liberate them from their stumps. Then I skid them to the sideyard with my 45-horsepower tractor and saw them into firebox-size rounds.

I find I am less effective at sharpening chain saws than at paying Vernon to do it. He sharpens them and sometimes I buy new ones. I buy gas for my saws, and oil and lubricants. They can fix my saws faster at Carolina Hardware and Fuel than I can myself, so I sometimes hire them to do tune-ups and repairs.

Once I have accumulated enough rounds, I need to split them and stack them under cover. My brother Mark often undertakes this as his exercise regimen.

To be successful I need to maintain two sheds: one for dry, seasoned wood and the other for wet. When I run out of wood I flag down Wilbur. He tops me back up for about $100 per pickup load, which is less than a face cord. I use about four face cords a season.

And once I have a shed full of seasoned wood, I need to get my chimney swept. I hire Charlie since he has all the gear and I don't. I feed the fire about a wheelbarrow full of wood per day and tote the ashes to the compost.

I figure heating with wood is about ten times as expensive as electricity if you factor in all costs. It's outrageously expensive. But it's renewable. And if I use just downed trees from my woodlot, I'm probably near carbon neutral. It's expensive as hell, but it's the right thing to do.

I find the fuel I make in my backyard is the same. I can conform to the $0.54 myth as long as I don't count labor or capital or other energy inputs. I would say my homemade fuel is way cheaper than heating with wood. It is probably only five times as expensive as petroleum diesel.

I used to incubate chickens with the neighbors. I managed a small flock for meat and eggs. I never did the math, but my guess is that my chicken meat was about $24 a pound. After all, I needed free-range, shaded enclosures that would protect the birds from weasels and possums and raccoons and coyotes and dogs and owls and chicken hawks. I rigged up a solar electric fence to keep the predation down. I fed the chickens organic compost and well water. The nice thing was that I felt safe using my own raw eggs in Mom's chocolate pie recipe, but I probably was paying a dollar an egg ....

It appears we will have a good garlic crop this year. Stacey planted it. We bought the organic seed at Wellspring. She built a raised bed for it. We had to protect it from the deer and the puppies, and we had to amend the soil it is growing in. I sure hope that when I calculate my cost per clove I come out cheaper than Food Lion. Don't tell me that organic vegetables grown in the backyard are ungodly expensive compared to store-bought.

As our biodiesel project has progressed, my wife Tami has been working on opening a local co-op grocery store called Chatham Marketplace. One night I spent some time considering how our basic economic structures are stacked against both biodiesel and whole foods. I addressed it this way in the blog:

Externalities

Last night I went to a fundraiser for Chatham Marketplace, the new co-op grocery store in Pittsboro that a bunch of people are trying to get off the ground. Tami is actively engaged in the effort, so I watch and participate with more than passing interest.

Organic produce and biodiesel share similar market plights. Both are in spaces where the "competition" has successfully externalized their costs. Big Food and Big Oil have figured out how to get society to bear their costs in ways that do not show up in the price of their products.

In the case of oil, these subsidies are painfully blatant. Diesel in North Carolina right now is about $1.59 a gallon. Of that, $0.49 is state and federal taxes. Not included in that $1.59 is the price of the fighter jets that escort tankers out of the gulf. Or the price of the current war in Iraq. Or the one before that. As a society we have agreed to pay those costs in other ways — through income taxes and volunteering our children for war.

Also not included in that $1.59 is the price of the health care necessary to treat the side effects of the fuel. We agree to ship our children off to school on diesel-powered buses, exposing them multiple times a week to toxic exhaust, and we find other ways to pay for the resultant respiratory problems.

If you want your asthmatic child to get health care in this country, it's probably necessary to have a full-time job with benefits. Better that than to meddle with the $1.59 price at the pump. Therein lies the subsidy. If oil had to pay its share for security, health effects and the environmental damage it causes, it would be outrageously expensive.

What is amazing is that food is in the same boat as oil. According to Cari Spring, in her book *When the Light Goes on: Understanding Energy,* the average meal comes from about 1,300 miles away from the table upon which it is served. That doesn't mean the food is expensive. We prefer to pay low prices for our food and to pay its real costs in other ways. To keep food prices low, farmers load their crops up with pesticides and petroleum-based fertilizers and spray them down with preservatives or perhaps irradiate them. Then they pack them onto refrigerated trucks and ship them off to markets on subsidized oil. Food is cheap because the tab for the children who are sickened by poisons is not paid at the plate. Neither is the price of the topsoil that is lost along the way.

The externalized costs of our food show up when you try to open a local marketplace. Since chemical inputs allow farmers to sell more of their perishable crop, thereby reducing the price per apple, the local apple from the organic grower in Silk Hope might cost more than the Alar-coated one from Washington state. This means poor Tami has a big education job on her hands if she wants to sell her organic wares.

One of the reasons I was so jazzed by my recent trip to California was that it appears the purveyors of B100 out there and in Colorado have successfully parted ways with the price of diesel. People are happily paying $3 a gallon at the pump. For $3 a gallon, we can easily make a renewable fuel out of vegetable oil — as long as our subsidies stay in place for the electricity, propane and methanol we use along the way ....

Here's hoping that Tami and the other folks of Chatham Marketplace can get enough people to give up their regular prices for food and switch to food that is locally produced, organic and less dependent on externalities.

After attending enough classes, discussions and speeches on biodiesel, I have arrived at the point of being able to handle almost all questions from the floor. Once, however, at the Southeast Sustainable Energy Expo in Asheville, a woman from the back row asked what the price of biodiesel would be if we subtracted all the petroleum subsidy that goes into planting, growing, harvesting and production and transporting biodiesel to market. I could not answer her question, but it was a good one and it casts another shadow over the attempt to read biodiesel economics.

Explaining the economics of biodiesel is a constant struggle at Piedmont Biofuels, as evidenced from this blog entry from the spring of 2005:

The Price of Fuel

The price of biodiesel just went up. Just as everyone is clamoring for cheap fuel. Our phones are ringing off the hook, not because of interest in renewables but because of pressure from petroleum pricing.

Once plugged into the classic laws of supply and demand, the price of biodiesel is painfully easy to explain, but for some reason those laws aren't supposed to apply to the average American's fuel costs.

The other day at lunch I was telling Evan I thought we should double our price from $3.50 to $7 a gallon. He was mortified and felt it was "counterintuitive."

But if we can't supply anyone with fuel for $3.50 a gallon, why not double the price and not supply anyone for $7?

When President Bush signed the Jobs Creation Act last fall, people kept asking us, "Now that there are tax credits in place, when is your price going to drop?" At the time, the answer was easy: "Nothing kicks in until January. I doubt you will see price reductions until then."

In January they were asking, "Now that you can get $1 a gallon off, when are we going to see that at the pump?" That answer was trickier. The IRS had not yet interpreted the new law, and producers were not going to cut prices until they saw revised forms and rulings on how the tax credit would impact their operations. We told people they might see some price reductions in the spring.

Now that the forsythia are fading and the dogwoods are starting their show, people are insisting: "The price of petroleum is outrageous. When are we going to see cheaper biodiesel?" And today our answer is that the price just went up. We are holding our selling price at $3.50 a gallon and we do have some fuel on hand, but forget about making any money by selling it at that price.

Those on the edge of the industry are outraged. They mention conspiracy and claim our suppliers are "greedy bastards," and they would love to cite their constitutional right to cheap fuel. Too bad the founding fathers left that out. My understanding is that they felt markets were an efficient way to deliver goods to those in need.

North Carolina is the fifth-largest biodiesel consuming state in the country. Its state contract stipulates that the fuel sold here will come from soy. The contract has been awarded to World Energy in the west, and Potter Oil in the east. We are in the middle and tend to buy from World Energy. World Energy cannot ship non-soy-based fuel into North Carolina, since most of their customers are buying on state contract. Potter Oil gets a lot of its supply from West Central Soy, which has customers on allocation right now.

Apart from an idiotic state contract that really should be feedstock neutral, why are there shortages of biodiesel right now? The answer comes back to tax credits, the price of petroleum and government policy. Some suppliers, including West Central Soy, have passed along a $1 a gallon savings to their customers. In some instances that has put their price at the rack (the terminal where fuel is wholesaled) lower than the price for diesel fuel.

However, diesel fuel is high right now based not only on $57 a barrel crude on world markets but also on increasing demand from China and India. Their economies have a greater dependency on diesel fuel in the first place, and as they grow they are competing with the US for fuel.

A lot of the costs in the production of biodiesel come from petroleum. The boats that import soybeans from

Brazil are powered by petroleum, as are the trains and tank trucks that deliver B100 to market. I've even seen grassroots biodiesel operations use petroleum to get their work done.

Where biodiesel is cheaper than petroleum, it gets snapped up by the general public rather than by customers dedicated to biodiesel. Construction fleets, cement makers and transportation intensive industries of all types — the same ones that previously were afraid to let B2 touch their precious engines — now are buying biodiesel to reduce their fuel costs.

Since I am quick to blame government policy for setting the price of energy, I should say that progressive legislation in Minnesota and Illinois has contributed mightily to demand. Minnesota has mandated biodiesel usage in state vehicles and Illinois has declared a tax holiday.

With all this demand, biodiesel is in shortage mode.

So when will the price of biodiesel fall? Simple. When there is a surplus. Biodiesel pricing will come down when petroleum demand eases (if it ever does) or when production capacity in this country increases. Before people freak out about their God-given right to cheap fuel, they need to realize that this is how the industry is birthed. One of the reasons Europe has a thriving biodiesel industry while ours is merely fledgling is that governments there used policy to make biodiesel more economical than petroleum. We need to do the same.

And as for production capacity? We're working on it.

eighteen

Microenterprise

I sometimes daydream about life in the Wild West. My vision includes riding into a small town and deciding what my job might be. Surely the decision would be based in part on whatever good or service the town lacked. If they needed a blacksmith, I might open a smithy. If the village lacked a barber, perhaps I would open a barbershop. If it needed a shoemaker, I might decide to make shoes. In the daydream I am unencumbered by lack of skill sets and can become anything I want. Sometimes biodiesel feels like the Wild West.

Today commercial biodiesel is not labor intensive. It is possible to assemble a highly automated process that will produce millions of gallons with fewer than a dozen hands on deck. As is the case with any industry motivated by greed rather than sustainability, commercial biodiesel has successfully replaced expensive humans with inexpensive capital. Most backyard and small-producer operations are not very capital intensive and therefore can be operated on an "occasional" basis. With less than $10,000 of investment, Piedmont Biofuels doesn't worry about sitting idle for a few days — as long as everybody's fuel tank is full. Conversely, when there are millions of dollars involved, often with an interest meter ticking or an expectation of return on investment, it makes more sense for the equipment to run 24/7.

In between the Wild West daydream and the capital-intensive commercial producer, I wonder if there might be meaningful employment to be found if we break small-scale biodiesel production into a handful of labor-intensive enterprises. I took a stab at an answer in Energy Blog and it went something like this:

Biodiesel As Jobs

Small Business #1: Waste Vegetable Oil Collection
This would require a pumper truck that can suck semi-solid veggie, chunks and all, out of a cold Dumpster. It also would require the ability to put a vacuum pump on a 55-gallon drum for grease removal. Since each full 55-gallon drum weighs about 400 pounds, this business would need to be able to easily lift these drums around. A Tommy Lift on the back of a heavy-duty pickup would

suffice. The business also might require some Dumpsters and storage containers.

The market: Local restaurateurs who would be interested in having their waste stream serve a higher purpose and a small local biodiesel plant that could offer $0.07 a pound for used veggie. At roughly eight pounds per gallon (including water content), a collector could fetch $0.56 a gallon for waste vegetable oil.

The competition: The rendering industry currently owns this business and recycles the waste veggie into animal food and makeup. The rendering industry is teetering on the edge of biodiesel production, with firms like Griffin Industries already producing fuel, but the business has consolidated to the point that the small veggie user has been forgotten. Individual restaurateurs also have largely been forgotten. The rendering companies are so "efficient" in collecting millions of pounds per week from giant producers that their customer relationships in Research Triangle Park of North Carolina are on the rocks. Our Dumpsters are flowing over now, which is why backyarders are so commonly invited to "help yourselves" to the used veggie resource.

Small Business #2: Straight Vegetable Oil Conversions
Whether using single-tank Elsbett or Neoteric kits or dual-tank Greasel or Greasecar affairs, there could be a small business in converting diesel cars and trucks to run on straight vegetable oil. Selling the tanks, radiators, solenoids and special fuel lines would be an obvious

sideline that would dovetail with the expertise. This business would require a solid knowledge of diesel technology and an automotive shop with a lift.

The market: Folks who want to run on SVO but don't have the courage or expertise to add a fuel system to their vehicles and would readily pay someone else to do it. At $300 per installation, installing one per week would be a $15,000-a-year gig. Not a bad part-time job. Those who know what they are doing can do a conversion in less than a day.

The competition: Do-it-yourselfers. Combining this business with grease collection to sell dried, filtered veggie to conversion customers would be a plus. The collector could fetch $1 a gallon or more from those who want to run on straight vegetable oil. Apparently there is a business in upstate New York called Liquid Solar that currently is doing exactly this: selling dried, filtered waste vegetable oil to customers who have converted their vehicles.

Small Business #3: Diesel mechanic

Someone is needed who is trained on both diesels and biofuels and can offer some comfort and assurance to the non-diesel-buying public. I once met a fellow in Asheville who billed himself as an "organic mechanic." He was helping customers green-up their vehicles in any way possible — from correct tire pressures to improve mileage to biobased lubricants to biodiesel. My guess is that such a line would be an ideal meal ticket for the future.

The market: People who care about their relation-ship to sustainable transportation but do not have the inclination to "do it themselves."

The competition: None.

Small Business #4: Used Diesel Car Lot
This business would create a source of vehicles that run on biofuels.

The market: While it is true that the new Jeeps ship-ping out of Detroit are leaving the production line with a factory fill of B5, the auto industry has a long way to go. Today 1 percent of America's driving public is in diesels, unlike Europe, with an adoption rate as high as 50 percent. There is an ever-increasing population in search of affordable diesel vehicles. The demand is met online and through word-of-mouth, and the absence of any knowledgeable retailers leaves a gaping hole that someone could fill.

The competition: None. Try to buy a used diesel at CarMax sometime.

Small Business #5: Tanker Truck for B100 Distribution
The grassroots community is full of tales of co-ops being formed, becoming successful and being wiped out by petroleum marketers who see the opportunity to jump in and meet demand. Increased demand is a good thing. A used tank truck for B100 distribution is about the same price as a roll-off tank with dispenser, yet the truck has more flexibility and can easily be

shipped off to another emerging B100 market once traditional fuel marketers enter the game.

The market: Individual consumers who want to buy B100. School bus fleets that want to self-blend.

The competition: None yet. This window may close as the community of B100 users grows, but there is plenty of opportunity at present.

I believe the list can go on and on. I haven't even scratched the surface of bioheat. Home heating fuel, after all, is simply diesel fuel that is dyed a different color for tax purposes. While all our work has been focused on transportation, there is an entire world of heaters and pumps waiting, and the jobs around the edges of that space could be similar. Another job that can easily be created by biodiesel is that of tour guide. I once tackled this in the blog:

Agritourism

The other day I piled into Tami's Jetta with a bunch of folks from the College and from Chatham Marketplace and drove to Lumberton, NC, to get acquainted with the Golden Leaf Foundation. They were created from money that came North Carolina's way via the tobacco settlement and they are all about helping tobacco-dependent, economically depressed regions to stimulate their economies.

They have funded cold storage and organic growers' co-ops and biodiesel operations in the past, and I figured it wouldn't hurt to learn more about them. One of the buzzwords they like is "agritourism."

A couple of weeks ago some guy on the Biofuels list wanted to visit a backyard operation and I invited him by "on list." Oops. I made a little joke about a "Biodiesel Bed and Breakfast" and the enquiries piled in.

We give a lot of refinery tours at Piedmont Biofuels — so many, in fact, that we have set a time, Sundays at one o'clock, to try to combine groups so that we can spend less time giving tours. Most of our groups are local, but having opened my mouth on the list I have been sending out URLs to visitors from afar. I've basically found myself in the agritourism business.

I was working with Neha, from our local Convention and Visitors' Bureau, on another topic and jokingly mentioned the number of queries I've had. She immediately sent out an "all points bulletin" and now the innkeepers are getting in touch.

Thursday I sent out the URL to Celebrity Dairy to four people: one from Florida, one from South Carolina, one from Virginia, one from Georgia, all of them into grassroots biodiesel, all of them well-informed. Celebrity Dairy is a working goat farm and bed and breakfast. The fellow who runs Celebrity Dairy is a graduate of the biofuels program, as are a couple of his employees. It dawned on me that if we have heads that require beds we should be directing them to Celebrity Dairy.

Friday I gave a tour to a fellow from Florida. He was headed to Celebrity Dairy, coming in from vacation in Richmond. We spent two hours together at the refinery on Friday morning. He is an enthusiast who wants to start an operation in his hometown. And at the end of the tour, he tossed twenty bucks into the pot.

It was a glimpse of a new self-image for me. I'm a $10-an-hour tour guide. I'll take that. It suits me. Maybe we should get a collection plate for the refinery so we can get paid to evangelize biodiesel.

I think it would make a nice tourism package. Pull into a Chatham County bed and breakfast and spend Friday night at the General Store Café in Pittsboro dining on local fare from local farmers and listening to live music. Come out to the refinery Saturday morning for a tour or to make a big batch of fuel and spend Saturday afternoon lollygagging around the myriad artist studios in the area. Then back to the General Store for a rocking Saturday night and light out Sunday morning for home with a working knowledge of home-brewed biodiesel, and goats, and some locally made pots or paintings. It sounds like a delightful weekend to me.

My brother Glen has the same "tourism" issue with his wind turbine. He could give tours until the cows come home and he tries to consolidate visitors to save time. We both live in rural areas with broad tourist appeal, and neither of us has had the brains to factor agritourism into our thinking about renewable energy.

nineteen

Fuel Distribution

It's still difficult to get biodiesel in North Carolina. Potter Oil, a small family-owned fuel distribution business in Aurora, covers one half of the state. To get to their place, simply drive to the coast and board a ferry. The other half is covered by World Energy, which keeps a steam-jacketed rail car full of B100 in Charlotte. Both groups have done a terrific job of education and outreach on biodiesel in this state.

North Carolina consumes more than the national average of biodiesel. I know of four B20 pumps that are open to the public and three pumps on the coast that are selling B5. My best guess for biodiesel consumption in the state for 2003 was around 750,000 gallons, from a national total of 25 million gallons.

But that is in petroleum blends. If you want to fill up on B100 you have to make it yourself, visit our refinery when we have fuel for sale or visit Carolina Biodiesel in Orange County. We have been attempting to remedy this situation as long as we have been in biodiesel. We purchased a little tank truck that can haul 1,500 gallons and we drive three hours to Charlotte to fill it up. Getting the tank truck on the road has been no mean feat, as I indicated in the blog:

Keeping Our Powder Dry

Our tanker has been making the rounds — not to paying co-op customers but to expensive stops along the way. First stop: Chatham Alignment, the local diesel expert. Second stop: Noble Oil Services for a tank and hose cleaning. Next stop: Maaco Auto Painting. We tried to find a local paint shop in Chatham, but ended up at a chain in Durham.

We feel it's important for the tanker to have a new, shiny look: a bright, newly painted green and yellow truck with our logo, our web address, our phone number and the tagline "Clean Renewable Fuels" emblazoned across the back.

But this is still a vision. Rachel's party at the College was well attended. The folks from T.S. Designs, Bo Lozoff and the Human Kindness Foundation crowd. All we sold was memberships to the co-op ($50 a year). Zero fuel. Our first press release was never sent because there still is no fuel to speak of. The Dodge is running on fumes. Next grease-run is tomorrow — and then we need to make a batch. There is a real possibility that I will be at a diesel pump with the Dodge for the first time since late January. I might not make it to finished fuel. I certainly won't make it to washed fuel. If I survive this drought it will be on unwashed. I may have to go down to the corner and make a donation to Operation Iraqi Freedom just to keep the Dodge on the road.

So in that respect these are discouraging days. I vanished into the mountains of North Carolina for a meeting with my brothers. We rented a perfect cabin on the banks of the Watauga in Valle Cruces outside of Boone.

Biofuels were one line item on the statement of a great weekend together. My brother Glen may be on the cusp of a massive investment in wind. He is an expert in portfolios — both financial and electrical — and his advice is that we "keep our powder dry" on other investments so that we will be ready when the big wind battle begins.

The fact that our tanker is not yet productive weighs on me. I have been preoccupied with other matters. Leif rocks. He has been working hard on the tanker. Rachel too. They are picking up my slack.

I have been discouraging membership sales for the past year because I have always felt they should be tied to fuel distribution. Members should be able to fill up. If members can't get fuel, something is not working.

I think this attitude is mediated somewhat by a conversation I had with girl Mark in Iowa. She's had a wild experience in the Bay Area of California, where she's seen co-op members without diesels, cars labeled "Powered by Vegetable Oil" that weren't even converted and everything else in between.

I've never wanted to be like that. I've always thought, "Fuel first, then drivers." Poor Cheryl, our first non-worker member, can't get fuel to drive across town. I sat next to her at Alternative Fuel Odyssey and she was making wisecracks about "fuel shortages." I'm glad we don't yet have a hundred members we can't serve.

I know the tanker will change all that. Memberships are trickling in. I watch Tami and Melissa and the other diligent folks of Chatham Marketplace sell memberships to the co-op grocery store. They are out there. They set up booths and promote themselves. I watch with awe.

At Piedmont Biofuels we are keeping our powder dry until the day we have fuel for sale.

We also have developed a small route distributing tanks to regular customers. It too got off to a rough start, as I indicated in the blog:

Tank Distribution Project

I'm not sure when it happened. I don't even remember who had the idea.

But at some point we decided to collect a bunch of tanks and distribute them to locations where our customers wanted fuel.

Rachel was on board. She was so enthusiastic about the idea that she published my cell phone number in the North Carolina Agricultural Review. And my cell phone rang. Thousands of gallons of storage capacity called in.

Some of the tanks were free. Some required a small payment. Some could be loaded on our truck and trailer. Some required that we do the loading. Some were full of unleaded gasoline or home heating oil or offroad diesel. And some were empty.

I took every call. ("It's got a small hole in the bottom, but that don't matter for what you're doing does it?") And I built a list based on some criteria. The tanks had to be in good shape, they had to be empty, they had to be local, they had to be free and they had to be loaded for $10 or less, which limited the scope of the project to about 6,000 gallons of storage capacity.

We are running all of this through Chessworks. Tuesday is picking up the tanks, bringing them into the shop, wire-brushing the rust off, cleaning them, priming them and painting them bright yellow with the Piedmont Biofuels logo.

And that's where we break down. The idea is that we will deliver thousands of gallons of capacity to our customers, follow around with our tank truck and distribute thousands of gallons of fuel. We could be just like propane or home heating oil.

What we have not yet solved is how the customer gets the fuel out of the tank. Each recycled tank is different. To install a barrel pump with a shaft that reaches of the bottom of each one is not simple. In the past week I have learned that there are a number of interfaces to consider. Some pump collars have detachable shims; some have butterfly nuts that are too large to turn on some bungs that really were not designed for nuts and bolts.

None of this has hit Leif's radar screen. We delivered our first 275-gallon tank with a barrel pump that has a shaft too short to suck fuel. It turns out that the shaft has custom threads. We delivered a nicely cleaned, nicely painted tank that can't be filled up with our fuel.

When I attempted to solve the problem, I ended up with a bunch of propane gas pipe fittings that do not fit with the project. When I went to the refinery to liberate a suitable pump, I discovered that there are different standards in play. There are fine-threaded barrel pumps with shafts made of tender steel and coarse-threaded barrel pumps with entirely different interfaces.

We need a standard, which is difficult when each tank is different. Some are oblong. Some are oval shaped. Some are fat cylindrical. Some are thin cylindrical. Some are elevated on stands. Some are on the ground. Each one will need a different pump installation.

Tonight Rachel and I wrestled one of our first tank deliveries into place. Barrel pump sucks from the bottom of the tank (custom shaft in place). Collar fits tank nicely and can be loosened as needed. But the customer still cannot draw off fuel because the hose we purchased does not fit the pump.

It's driving me crazy. We have more than a tanker's worth of capacity waiting in a queue for B100 and we cannot deliver. Some of our tank customers have been waiting since the spring.

And here is my trepidation. This week we had a complete stranger call looking for a tank of B100 for her farm near Silk Hope. She saw that we were exhibiting at Shakori Hills this weekend and went from "never heard of biodiesel" to "get me an onsite tank" after a brief survey of the net. Today I had a fellow call who had heard my interview on "The State of Things." He wants a 150-gallon elevated tank to fuel his 2005 Passat wagon and possibly a tractor.

Yikes. We have incoming capacity. Tuesday can pick up an unlimited number of abandoned tanks. And we have outbound capacity. Based on our tank backlog, the tanker could empty itself in a day. But we are cross-threading pump shafts in place because we are wholly unprepared for this demand.

In a blatant act of over-enthusiasm, Rachel applied for a grant from Triangle Clean Cities that would allow us to place 500-gallon tanks, with pumps and filters, around the Triangle. The grant was to cover six locations and it gave us one year to do it. We gave one tank to Carolina Biodiesel and have installed one at the refinery that has become a delight and was christened the "Tami Tank" in the blog:

Everybody's Full

I returned from a week on Topsail Island, unloaded bikes and kayaks and sandy towels and jumped in the Dodge to drive to a board meeting of Carolina Biodiesel Industries, 40 minutes away. I know I am running on empty. My new "B100 distance record" will have to wait, as I decide to fill up on petroleum to make it to dinner and the meeting.

But on the way I pass Leif, driving the Piedmont Biofuels tanker down the Moncure Road. It was a bright yellow beacon barreling toward me. I hit the brakes, cancel my dinner plans, do a U-turn and head for the refinery. When I pull in, Leif is giving a tour. Oneas is working his sunflower stand. The chickens are contentedly dining on grubs and cracked corn. And in my absence the place has been transformed.

Oneas has built a pair of paddocks to house the goats he has recently acquired. Leif has ripped the Princess Palace apart and has started constructing his temporary lodging. Doug has expanded his composting operation and it appears the vermiculture project is underway. It's a remarkable late summer evening. I interrupt Leif's tour to enquire about the status of homemade fuel. He shrugs me off and says, "Help yourself."

But when I check the usual fuel barrel (the one on its dolly that I now look into prior to pumping), it's empty. There's 12-volt pumping gear lying atop the wash tanks, but as near as I can tell, Wash 1 is empty. I interrupt Leif again for a 12-volt pumping lesson. He graciously leaves his charges talking with Oneas about his sorghum crop and walks over to the refinery.

"I'm 12-volt pump challenged," I explain.

"I understand that's by choice," Leif grumbles.

"It is by choice," I admit, "but I still could use a hand."

He studies the wash water remains in Wash 2 and says, "I guess we are out of fuel."

"So everybody is full?" I ask.

Leif grins. "Rachel is full, Scott is full," and on and on he goes.

"Everybody's full," he says. "I guess it's the first time ever."

We wander over to our elevated storage tank and to my delight Elaine has begun painting it with sunflowers. Chris, who has now returned to school from his summer stint with us, finished plumbing the unit, and it is sitting full, complete with pump, fuel meter, filter and "automatic shut off" dispenser. It's a project Ward started months ago that finally has reached fruition.

"It's a Tami Tank," I declare. "I could fill up here!"

Generally, filling up at the refinery can be a heavy, greasy, messy job. For years I've poured homemade biodiesel out of gas cans or out of carboys or barrel pumped it by hand, but this time I lean on the hood and chat with Leif as the Dodge is filling up with store-bought B100.

The nozzle stops dispensing fuel 28 7/8 gallons later, just like at a real gas station. Tami can now drive up in her Jetta, fill up in stiletto heels and not spill a drop.

It was fantastic. I drove off to my board meeting with my B100 record intact and the knowledge that "everybody's full."

———————————— ⌒ ————————————

After three years of work, there still are only two locations in the state where consumers can buy B100: us and Carolina Biodiesel, who for the time being are getting their fuel from us. Pathetic. The miniscule distribution infrastructure reflects the fact that we have been out of fuel repeatedly since we started this project.

Hopefully the day will come when there are enough consumers in the area that the petroleum marketers will notice us and put B100 in their stations.

twenty

The Policy Layer

The idea that the deck is stacked against renewables is not original. If biodiesel is twice the cost of petroleum, it is because petroleum gets massive societal subsidies. Electrons generated from wind are "more expensive" than those from a nuclear plant because wind turbine operators have to pay for their own grid, pay their own debt and pay their own way, while we as a society have decided to waive these costs for our utility companies. Energy from active solar costs four times as much as that from the grid, again because the grid gets so many free passes instead of paying its true cost. The policy layer dictates most of the cost of our energy.

None of this is radical thinking. It's more or less accepted truth. The question then becomes how much effort can we expend on changing the policy layer. I have to say that my own excursions into this area have been bruising and largely unproductive and have left me shy of the entire process. Here is a pair of blog entries that I published along the way:

Making Legislation

I would rather write uplifting blog entries. Because the legislative process has been frustrating, slow and unproductive, I haven't seen the point of covering it. Up until this morning, I felt as if all the energy I have put into legislation has been nothing more than a waste of precious fuel.

My brushes with legislative change have come from my involvement with the legislative subcommittee of the Triangle Clean Cities Coalition. That group is populated by a bunch of folks I admire and respect. My idea of a tax holiday for biodiesel in North Carolina never got much traction with them, since most of their constituents are governmental and don't pay road taxes anyway.

They are into blends. I speak for B100. They are into fleets. I am involved with consumers. They know a lot about all sorts of alternative fuels. I don't much about anything but biodiesel.

My legislative efforts haven't amounted to much more than a cup of warm spit. I have Joe Hackney's ear. He's the House majority leader. I have Jim Black's ear.

He's the House Speaker. But my concept of a tax holiday for biodiesel never got anywhere, and I have been investing my energy elsewhere.

In Iowa this week, with the kids in tow at my daughter's high school graduation, I get an e-mail with a proposal from Rep. Joe Tolson for House Bill 1636. It infuriates me. It defines biodiesel as fuel made from pure virgin soy, leaving out sources such as yellow grease and rendering and mustard seed. If North Carolina is the Saudi Arabia of biomass, why would we limit our legislation to pure virgin soy — a crop for which our state is on the cusp of a deficit? If we were to get a 13-million-gallon biodiesel plant open (the proposed size of the one the Grain Growers are working on), we would need to import the soy oil or beans to make the fuel.

I was so furious when I saw the proposed legislation, I sat down to write a blog entry. I took the bill and ripped it apart paragraph by paragraph in a very sarcastic, condescending and indignant way. Then I stopped myself. Before hitting the "Publish" button, I read the entry aloud to Tami and the boys. It put both the boys to sleep, which always is a good thing, but it drew cautionary remarks from Tami.

The bill says nothing about conservation. In fact, it sets "minimums" that consumers must hit in order to qualify. It basically says, "Drive more to qualify. Don't worry — it's renewable."

My first essay was full of wrath and bitterness toward all parties involved. I ripped into Big Soy for requiring virgin product, Big Government for requiring minimums, the State Energy Office for neglecting conservation. You name it, I blasted it. I was seething.

Then I thought about Bo Lozoff's observation that biodiesel is a "movement," and that movements work best if we all work together. Tami was right. Blasting the crap out of idiotic legislation might drive a wedge into the movement. I don't want that.

So today I composed a very clinical, "disappointed" e-mail to the stakeholders involved. I sent it not only to the people of the Clean Cities subcommittee but also to Leif and Rachel and my brother Glen — people I can trust to calm me down. It was long, sanguine and rational, with only a hint of "I'll not only not support this but I'll take whatever political capital I have to defeat it." Fortunately my computer was having difficulties and the mail was never sent.

I could not drop my rage over this issue, so I found Anne Tazewell's number, called her on my cell phone and left a message. Anne runs Triangle Clean Cities. We have to work to get along with one another, but we generally do and we have a mutual respect despite the fact that I have largely written off all my "lobbying" efforts.

Zafer and I were having breakfast at the Audubon Restaurant in Ames, Iowa. We were marveling that on the wall there was a print of Audubon's "Cedar Birds" and in the birch tree outside there was a flock of cedar waxwings. We were discussing how the names of birds could change over time — an osprey was once a fish hawk — when Anne returned my call.

I was delighted to hear that she was furious too. House Bill 1636 had caught her by surprise and she could enumerate its idiocies. She didn't need a scathing blog entry or a snarky e-mail from me. She gets it. Which means the State Energy Office gets it. Which means there is hope. Which means we just have to figure out who the numbskulls are who drafted the legislation. Probably Big Soy, Big Oil and Big Government.

Apparently this is what happens when you set about making legislation. If I can figure out a way to check my fury at the door, I might be able to counter the opponents of reason with Big Sustainability.

Big Sustainability

My parents are visiting from Guelph, Ontario, this weekend and we had a great talk about the Ontario electricity market. My father and mother have an energy bill that is about 25 percent less than the provincial average, and they use no alternative energy whatsoever. What they do is conserve.

In Ontario the market for electricity has been capped by the past two governments. Each ratepayer in Ontario has a line item on his or her monthly bill that reads: "$4 Debt Reduction." In essence, the government of

Ontario has decided to pass along the cost of electricity from nuclear plants by collecting $4 a month from everyone on the grid. Once the debt is paid off, they can "deregulate" and have a true market.

North Carolina has no true market for electricity. We have stranded all our costs in projects like Shearon Harris, a giant nuclear plant around the corner, and we are letting them ride. Because the cost per kilowatt-hour of generation is incalculable, we'll never know if the bonds we passed were a good investment.

It strikes me that my casual conversation tonight with my parents and my last blog entry on making legislation come down to the same thing: there is no constituency for sustainability.

Surely the concept of Big Sustainability is misguided at best. Anyone who has read E.F. Schumacher's *Small is Beautiful* can tell you that. And when I think about it, everything I admire is "small," or local, which often means the same thing.

Small business is where the jobs are created. I like that. Buying locally keeps small producers afloat. I like that. I believe there is a chance that "small" might lie at the heart of sustainability. And yet when it comes to influencing governments, "small" has no clout at all.

Take the solar industry. It's been flourishing since the 1970s and yet its penetration rate is miniscule. That doesn't make it bad; I love the folks involved in solar. Unlike "biodiesel," "solar" is in the vocabulary of Joe Six Pack but he doesn't use either to meet his energy needs.

The folks I know in either solar thermal installation or photovoltaics would readily unite with biodiesel — or at least groove on it. And so would the people in microhydro and wind. The problem is that even if all the alternative energy folks put together a single voice it would not add up to Big Sustainability, which might get noticed by our elected officials.

I can see it now. The legislators gathered around in a backroom meeting, the air thick with the cigar smoke of Big Tobacco:

"Big Oil won't fight that. They'll get the credits too."

"I can't see any trouble from the farm lobby. The people over at Big Corn say that as long as we keep ethanol in they'll be happy, and of course the bill was written by Big Soy."

"Big Government won't oppose it as long as it's easy to collect and gives them something to spend."

"What about Big Sustainability?"

The room falls silent. "You know we have no chance of passing this unless it gets blessed by Big Sustainability."

Still no answer. "I'm not putting a bill out there unless Big Sustainability is going to turn a blind eye to it. I'll tell you that much right now. I'm not gonna see my effectiveness rating drop because I didn't have the sense to consult with Big Sustainability."

Nice daydream. Time to get back to work.

I was so discouraged about my travels through the policy layer that I decided to abandon the effort altogether. I'm not diplomatic enough. Not patient enough. And I find myself unwilling to abide sheer stupidity.

Rather than attempt to even the playing field for my beloved renewable fuel, I decided to persevere with a total separation between biodiesel and petroleum. "Want the cheap stuff? That's down on the street corner. What I have to sell is twice the money ...."

But just when I was resigned to abandoning all work in the policy layer, I had a dream, which I wrote about in the blog:

Approaching the Government

Last night I had a vivid dream that convinced me we are not doing enough. I awoke with a start, horrified by the state of things in North Carolina, which essentially is a biofuel backwater.

In the dream I was on my way to Raleigh to initiate conversations with the powers that be. I awoke with a renewed sense of urgency, and I called Richard Hardraker for advice. He's the guy who shepherded through the NC Renewable Energy Tax Credits and he's a force behind NC Green Power. My understanding is that he was appointed by the governor to guide the state on energy issues. On the phone he was his usual modest, knowledgeable self who had the patience to listen to me rant on about what we should be doing legislatively to get biodiesel on North Carolina's map.

Next morning I cleared my calendar and went off to Clean Cities. The more I considered the options available, the more I focused on a simple, concise goal: no road taxes on biodiesel. Today about $0.49 on each gallon of diesel goes to state and federal taxes and $0.24 of that is for the North Carolina road tax.

Simple. Give biodiesel a tax holiday. Doing so would not impact government revenues. They collect virtually no road tax for biodiesel now, since hardly anyone runs on the stuff. Doing so would be easy to administer. There would be no tax to collect.

While some people may be drawn to backyard biodiesel as a method of tax evasion, I am not. At the same time, if I run 100 gallons of straight veggie through my system, where do I send my $24 check to pay my share of North Carolina road taxes?

Road taxes seem to me fair and equitable. If I use the road, I pay my share for its upkeep. The more I use it, the more I pay. No objection there. But making biodiesel tax-free would reduce the uncertainty and doubt that currently surround homemade fuel in this state. It would be nice if backyarders had to concern themselves only with the EPA and the IRS and could take state "revenuers" off their list.

Alternative fuel legislation in North Carolina has been led for the past two sessions by a bill sponsored by Rep. Joe Tolson to increase the fees collected on license plate renewals. By collecting an extra buck every time someone renews, the state would have a good pot of money to put into the development of renewable fuels. Good idea.

The bill failed the first time because legislators from rural districts viewed it as a tax to clean up urban problems. Alternative fuels reduce emissions. Reducing emissions helps with air quality. Since air quality is largely an urban problem, the logic went, get your slimy paws off my rural constituents. I can almost hear the detractors in my head: "Our air ain't so bad down here. Why should we have to pay an extra dollar?"

Dave Ronco from Wake Technical proposed what I thought was a brilliant solution to this rural-urban objection. In North Carolina, some counties require emission controls testing for safety sticker renewals and others do not. As our air becomes increasingly polluted, more counties are included. His idea was to make the increased license fees follow the emission controls counties. Another good idea.

But the bill failed again. Apparently some of the opposition came from the NC Petroleum Marketing Board. They are opposed to compressed natural gas and propane and even ethanol. They don't sell the first two, and ethanol is so hydrophilic (water loving) that they don't have the infrastructure for its proper storage and handling. Assuming that we must defend their greed at all costs, we surely should not start a fund for alternative fuel development.

Recently they conceded that they would drop their opposition to the Tolson bill if the fund focused only on biodiesel. Great news. They can make a buck blending biodiesel into their petroleum products, so they will acquiesce. Whatever we do, let's make sure we support the petroleum industry with our legislation.

My thinking on the Tolson bill has changed some. At the end of the day, people don't like the idea of government amassing money to do anything. There is a prevailing sentiment that says, "I can do a better job of handling my money than the government can," and as long as this attitude persists, the idea of collecting an extra buck from everyone may be dead in the water.

At the Clean Cities legislative subcommittee meeting, Tolson's bill was discussed at length. Maybe if it were $0.50 instead of a dollar? I proposed my idea: no taxes on biodiesel. It was passed around the table again and again. Anne Tazewell was concerned that it wouldn't help government. Government is the big user of biodiesel in our state and it doesn't pay road taxes.

Kurt from the Solar Center liked the idea. There were a couple of engineering professors from Appalachian State there who didn't really know what biodiesel was. They were ethanol guys. Some folks from the State Energy Office were there. They seemed to know their way around the legislature but you could put their knowledge of biodiesel in a thimble.

"But people can't run on B100 can they?"

"Yes, they can. It's how I got to this meeting. Let's make it tax-free."

"But doesn't it void engine warranty?"

"No, it doesn't. Let's make it tax-free."

"But what about the federal taxes?"

"Not our problem. Let's make it tax-free."

"But what about people who run B20?"

"That would make twenty percent of their fuel tax-free."

At the end of the day, Anne's idea prevailed. It involved some sort of tax credit for those who are blending biodiesel into their petroleum products. That's OK. Clean Cities is fighting the good fight, and they know more about starting conversations with the legislature than I do.

With time, the sense of urgency passed. I took my family to Florida for a week and forgot about the issue entirely. I returned to e-mails and voice mails from a bunch of legislators and even one from the governor's office. My fire was relit.

I called Anne to let her know I was going to push on with my idea of the tax holiday for biodiesel in North Carolina. She was supportive and encouraging, and although she couldn't give me an official blessing she was glad to be in the loop. Helping the government can be her thing.

I want to see the public using B100. And I guess I'm simply not interested in blenders or in agencies with big fleets. So I will push on. Of all those at the table at the Clean Cities subcommittee meeting, I am the only one who is legally allowed to lobby. I have no experience. I tend to be hotheaded and impatient. But I am passionate about biodiesel. Over at Piedmont Biofuels, Leif has an immediate opening for a lobbyist. I think I'll go over there, take the "Help Wanted" sign out of the window and apply.

I had no luck with the tax holiday idea with the Clean Cities folks. They concocted a "blenders' tax credit" that would go to petroleum blenders in a vast, convoluted formula I could not understand. Instead of fighting about it (although I did have some heated conversations with Anne Tazewell), I decided to step out of the policy layer and go back to fuel making where I belonged. Anne probably has done more for biodiesel in North Carolina than anybody else and I adore her, though we scrap mightily from time to time.

Our latest head-to-head was over the pricing of B100 at the pump. She has taken my seat on the board of Carolina Biodiesel, and I objected to one of their posts about how they are going to lead the fight to make biodiesel cheaper than petroleum. She lives in the subsidy layer. She gives money to people to buy down the price difference between the two fuels. This is an area where I see no hope. The only way to make biodiesel cheaper than petroleum is to externalize its costs as petroleum has done, but building up a subsidy infrastructure equal to petroleum's is a lost cause. My thinking is

"Forget it." Let petroleum prices come to us. This will happen.

Piedmont Biofuels has benefited from some of Anne's subsidies. She makes money available for the zoo, and the zoo orders their B100 from us. But while it is nice for us to get big subsidized orders, at the end of the day it is simply PR quantities of fuel that will not continue to be purchased after the subsidy runs out.

Anne wants to subsidize (which is her job, after all) so that the two fuels can be equals. In my mind they are not equals and need to be treated differently. As she and I go round and round, our engagements with the policy layer continue, and I wrote about them in the blog:

Alternative Energy Economic Development Day

This was a big day. Piedmont Biofuels was invited to participate in an economic development summit at the NC legislature. I went to the refinery at 6 a.m. to pick up the tanker. Our goal was to park it on the lawn of the legislature as a giant prop for all to see. It would not start and it broke my heart.

I piled into the Dodge and headed to Raleigh anyway. Rain splashed on Highway 64 across Jordan Lake, and I hoped it would pour all day so the absence of our wonderful tanker would be less noticeable. By the time I arrived in Raleigh, the rain had stopped. As exhibitors arrived I apologized for the absence of the tanker and helped people carry their displays up the outdoor stairway.

Most of our toting involved walking past a giant Hummer with NC legislature plates. Some representative clearly was proud of his gas pig and had parked it front and center. The event was sponsored in part by the North Carolina Sustainable Energy Association. Their executive director, Mark Ginsberg, is a trade-show warrior. You will find him setting up display materials at every conference in the state.

When Leif and Rachel arrived, we set up our booth next to the Appalachian State Energy Office. There was an amazing professor who had written the state's energy plan and knew a great deal about energy policy. Everybody in alternative energy was there. On the green between the Legislative Office Building and the Department of Public Instruction was a series of tents. Smithfield, the giant hog producer, was smoking pigs in preparation for serving lunch to hundreds. The absence of the tanker overwhelmed me. It was our chance to show off, our chance for biodiesel to leave a mark.

I headed for Joe Hackney's office and one of his aides walked me over to an adjoining building and pulled him out of a finance committee meeting to speak with me. I was blown away. He explained that the NC Department of Transportation had weighed in with opposition to the idea of a tax holiday for biodiesel. They countered with a proposal that would exempt fuels made from waste vegetable oil. Joe seemed to like that.

I went back to the booth and begged Rachel and Leif to get the tanker onto the scene. I felt the problem was electrical and they concurred. They headed to the refinery to save the day. Along the way they called Tami, who took a left turn in her daily routine, headed to the refinery and dug into the tanker in her office clothes. She called in a part number to Rachel, who then was able to order the correct batteries.

I spent the morning talking to Progress Energy, the Farm Bureau, a lobbyist from the NC Solar Center and the folks from Carolina Green Energy. It was amazing. The summit was sparsely attended but those who were there knew the issues intimately.

Rachel and Leif roared in with the tanker. We parked it beneath a tree on a massive green in front of the legislative building, right next to the "pig pickin'." It was stunning. Everyone going to lunch had to notice it. Hundreds of folks. It was fantastic.

I talked to Joe Tolson at lunch. He's the sponsor of House Bill 1636, which Clean Cities is backing. We talked about feedstock neutrality. I talked to Alex Hobbs. Word on the street is that his group believes the ASTM biodiesel specification favors soy. If you ask me, it favors petroleum.

After lunch we all enjoyed hearing from a Pennsylvania dairy farmer who was creating electricity from manure and selling it back to the grid. In a state where net metering is still illegal, the questions from the floor were not that impressive.

Leif got the call from Capital City Police. Rachel was grinning as we headed down in the elevator. We emerged in our new mustard-yellow golf shirts, sporting the Piedmont Biofuels logo, and met the friendly officers, who wanted to see our parking permit. Of course there was no permit. We had been invited. "You can't light it with a match," I said. "Some people say you can drink the stuff."

They weren't impressed, and Leif removed the tanker. Rachel and I broke down the show booth and we all high-tailed it out of Raleigh. She and Leif headed for a meeting of the steering committee of Triangle Clean Cities and I drove deep into Orange County for a board meeting of Carolina Biodiesel Industries.

I had been working on biodiesel for about 11 hours straight when someone remarked about the different energy levels between the private sector and the not-for-profits. He was complaining that no one in the not-for-profit space has the passion to "get people into the store." I wondered if Leif's and Rachel's brains were hurting as much as mine. Combined, we had put in 36 hours of passion and I knew that all of us would still have a ton of e-mail to answer later.

It was a big day. I came home and did headstands with Tami and the kids, who had just returned from a yoga class ....

In the winter of 2004 I found myself reenamored of the effort to affect the policy layer and tossed the "tax holiday" idea back into the ring. This time it received a much better reception, including a broad endorsement by our local Clean Cities crowd.

I have to thank the National Biodiesel Board for renewing my enthusiasm for changing legislation. With "supply side" and "demand side" in my vocabulary from the 2004 NBB conference, I was reinvigorated for another brush with democracy. Several ideas found legs at the same time. I addressed them all in the blog:

Lobbying Time Again

During the last short session of the NC Legislature, I ended up bruised and battered by a scuffle with the renewables establishment over a tax holiday for biodiesel. Things have quietly reconvened and it's lobbying time again. This year there are five primary ideas floating around.

The first is "to set a renewable fuel use standard for state fleets." Minnesota has done this successfully and it is a boon for both ethanol and biodiesel. With credit being offered for the deployment of hybrids, this strategy would not only "green up" the NC fleets but also create an attractive market for fuel makers.

The word that comes to mind is "mandate." I have no trouble with sensible mandates, but I'm not sure I want to try to sell one to the NC public. My guess is that the term has been cut from the true lobbyist's vocabulary.

The second strategy afloat is "to sell Energy Policy Act (EPACT) credits to generate funds for alternative fuel use." This is a fantastic idea. State governments accumulate EPACT credits when they acquire alternative fuel vehicles and burn alternative fuels in them. Missouri just sold 1,000 EPACT credits to a utility in California for a million bucks.

This idea would take that million and throw it into a pot for alternative fuels. I would say it is a good idea even if the money simply goes into the general revenue fund. It doesn't matter to me if you pay down debt or offer raises to educators or endow an arts project. Selling EPACT credits would help the market develop, and a market for clean air is a good thing.

If we ever are going to come to terms with the true cost of energy use we have to get rid of externalities, and I think the marketplace is an excellent place to do that. We should be trading in carbon and clean air and everything else associated with our consumption patterns.

Idea number three, a no brainer, is "to eliminate the state motor fuels tax on biodiesel and ethanol." Whoever introduced this idea is a genius. It's not an original thought, but it would give the fuel a boost and would relieve the tax burden on home-brewers.

Idea number four, which has been around for a while, is "to raise motor vehicle registration fees to generate funds for public and private sector use of alternative fuel and advanced transportation technologies." This is Joe Tolson's perennial idea that never seems to get traction. If the state could grab an extra dollar or two every time someone renews their plates, there would be a huge fund to support alternative fuels. I think it's a great idea. But for simplicity's sake, let's just label it "tax and spend." Then let's kick back and wonder why it never passes in North Carolina.

The fifth and final idea in play right now is "tax credits for alternative fuel vehicles (AFV) and hybrid electric vehicles (HEV)." This one is hard to argue with, except I'm not sure it would do anything for the B100 Community. In order to benefit from tax credits, a tax liability is necessary, and those driving around in 25-year-old diesels aren't known for their tax liabilities. But it might appeal to some.

It appears that there is NC Senate support for all five ideas at this point and that some support from the House is now needed. So here we go again. I'm going to try my hand at drumming up some House support for these initiatives. We'll see if those who remember our last attempt return my calls.

twenty-one

The End of Oil

I gave up on the "End of Oil" argument a long time ago because I no longer found it useful in the biodiesel conversation. My introduction to it came from *Hubbert's Peak* by Ken Deffeyes. He's a crusty old geologist who has had an illustrious career at Princeton and before that worked for Shell Oil's research laboratory, which he claimed was the best earth science research organization on the planet at the time. He writes with authority, he is a magnetic speaker and I believe his notions of peak oil are accurate and compelling.

I find the same is true of Richard Heinberg, the author of *Powerdown* and *The Party's Over.* He clearly is well-informed on the subject and does a wonderful job of including fear in the conversation. In his marvelous book, *Powerdown: Options and Actions for a*

Post-Carbon World, Heinberg creates a striking image of being trapped on a leaky raft full of people partying, gambling and carrying on, oblivious to the fact that the raft is about to sink. He finds a group on one corner of the raft that is carefully mending it to make it water-tight and durable, but since they are using materials collected from another part of the raft their efforts are unsustainable.

The "End of Oil" controversy is captured in Gregory Green's *The End of Suburbia,* in which various authors discuss the impending global shortage of both oil and natural gas. And of course the Internet is ripe with well-meaning harbingers of doom.

In the early days of Energy Blog, I sometimes would respond to readers' posts, and I once tackled the peak oil argument this way:

Hubbert's Peak

Stacey writes: "I heard that a few scientists at Sweden's Uppsala University have predicted that the previous forecast of oil and gas supplies running out by 2040 should be revised to 2010. Is this true among your intellectual circle and, if so, what will we do in seven short years to have minimal energy needs met? If this is true, how do we plan for this as individuals and as a community?"

I think anyone who has read *Hubbert's Peak* would shrug when reading the current forecast from Uppsala.

Most of us buy into Deffeyes' basic arguments about world oil reserves. He was the keynote speaker at NC Fuel Choice and I'll paraphrase his talk. First I should say that he is a short, round, gray-haired academic with a feisty and cantankerous delivery style. I found both his book and his talk to be brilliant and engaging, despite his "curmudgeonly" tendencies. He said something like this:

"The United States' oil reserves are short by one Kuwait. I call up my friends at the United State Geological Survey and say, where is it? Is it in Alabama? Where in this country have geologists not searched for oil? Where have they missed the mother lode? Tell me where it is and I'll go halvsies with you."

This is from a guy who nailed the largest oil find in New York. This is from a guy who has discovered oil reserves all over America. This is from a guy with 100 out of 100 successful oil wells and no dry holes.

"Unless we include Iraq," he says. "If we include Iraq as the 51st state, suddenly our reserve forecasts balance."

On the same tack, Deffeyes claims that world oil reserves are short by one Middle East. The only part of the planet he feels is underexplored is the South China Sea, which he describes as "three rocks sticking out of the ocean with a different country's destroyer circling each."

I don't suppose I speak for all of us in Piedmont Biofuels, but I think there is a broad acceptance among us of the imminent disappearance of fossil fuels. I also think there is not much stress over the matter because there is an understanding that humans will be able to innovate their way out of or around the impending crisis. I fall back on the inspiration of Amory Lovins, who argues that we are the ones who should innovate and offer solutions rather than handing the next generation a world in which innovation is urgent.

I have four children. They figure frequently in the Energy Class and in the blog. Four unsustainable children. Each of them is amazing and worth investing a lifetime in. I look at them, including my teenage daughter, Jessalyn, who likes to lie on the couch and announce, "Dad, I will do whatever you want today as long as it has nothing to do with the natural world," and wonder, "How can I make this work?"

So my basic answer to the researchers in Uppsala is, "What's new?" Stacey's question is much harder than that. What will we do in seven years to keep minimal energy needs met and how do we plan for the end of oil? I don't know the time frame, but I believe innovation will get our needs met. And the way to plan for the end of oil is to let innovation happen.

———————————— ⌇ ————————————

To me, fear is a rotten motivator no matter how masterfully it is presented. The Bush-Cheney administration has successfully harnessed fear to get their policies through and to win reelection, but that doesn't make fear valuable in our current energy conversation.

When I was growing up, the predominant fear was about being sent to fight in Vietnam. When that fear subsided, the Cold War concept of mutually assured destruction took center stage. After the Peace Dividend there was a hiccup of fear surrounding the turn of the century with the Y2K fiasco. And now there is fear mongering surrounding Peak Oil.

Perhaps my generation functions best against a backdrop of fear. There certainly are many in my community who feel that a devastating oil shock and inevitable currency collapse would vindicate their voluntary simplicity. But my preference is to forget fear. Surely we must all be diligent about reducing our ecological footprints, and few people pay much attention to the message of Amory and Hunter Lovins about how we can conserve our way out of energy shortages. But as for being afraid? No, thanks.

My final abandonment of the "End of Oil" argument came one day at Chessworks when I engaged in conversation with a right-wing unemployed repo-man who was lying about the place and keeping a watchful eye on our biodiesel efforts.

I played the Hubbert's Peak card and he ignored it. "Hell, they got reserves in Saudi Arabia they ain't even discovered yet."

How do you argue against that? Instead of trying, I simply abandoned the effort to convince people that oil has peaked and that we will one day run out of our fossil inheritance. Of more use is the argument that biodiesel can be made in America and could offer manufacturing jobs to an economy in desperate need.

It's a less scholarly approach but a whole lot more effective in the end.

twenty-two

The Road Ahead

Running engines on biofuels is old news. A diesel engine running on peanut oil was shown at the Paris Exposition in 1900. Rudolph Diesel himself once said something like: "We won't always have fossil fuels to power our engines." The petroleum industry seized upon his invention and recognized that it could happily burn a distillate that was just above asphalt in their waste stream. They called this "diesel fuel" and we have been spewing out its particulates ever since.

The nice thing about biofuels is that they are not dependent on nonrenewable sequestered carbon reserves. They are solar energy. The bean takes its energy from the sun; we crush the bean into vegetable oil and burn it in our diesel vehicles. It's carbon neutral. We have exhaust, like most other vehicles, but we are busy belching the same quantity of carbon into the atmosphere that the plant originally collected from the air.

If we are not interested in annexing Iraq and the rest of the Middle East, perhaps we should be considering fuels made from seed crops we can grow right here at home. At Piedmont Biofuels we do several things. We educate folks on how to break their addiction to fossil fuels — sort of a rehab counseling service. We convert diesel vehicles to run on straight vegetable oil. We lobby our government for a more favorable legislative environment for biodiesel. And we make the fuel.

Biodiesel is great fun. It's empowering. Nothing feels better than tooling down the highway with the knowledge that you are free. Free of Chevron. Free of Mobil. Free of George Bush. Free of the Saudis. Free of the whole sorry lot. I realize full well that hydrogen is the place to be, but I'm stuck on vegetable oil. It's here now. It works. It's renewable. It's sustainable. It smells good. It creates jobs in the United States. And there is no war required to get it.

The other day OPEC elected to curtail production and the markets took a big spike up. The petroleum markets, that is. The stock market took a dive. It got the "jitters." Biofuels don't have much impact on the stock market. In an economy that currently is decrying the loss of manufacturing jobs, biofuels don't even register. We are Sally Brown in the Peanuts comic strip, burying our heads in our hands to the caption: "He doesn't even know I exist."

How far can cleaner fuels for the same old engines take us? Not very far. If you ask me, how "far" we can go has nothing to do with our fuels or our engines. It has everything to do with us. Our habits. Our attitudes. Our understanding of sustainability. Want to be sustainable? Walk. Ride a bike. Take a bus. Get rid of whatever gas pig you happen to be driving right now.

Biofuels are a way of sticking our finger in the dike. Real solutions lie in a wholesale change in the way we interact with the planet. Real solutions involve the reinvention of local enterprise and the extinguishing of greed. But if you were walking by the dike and noticed a dripping leak, would you stop and put your finger in it?

You can build your own biofuel refinery in your garage. If everyone did that we would have a shot at upending our typical energy infrastructure and moving to a micronodal delivery system that might be borderline sustainable. And if we all made our own fuel out of vegetable derivatives, the feedback loop alone would mean that we'd drive significantly less and we'd get a keener understanding of sustainability. Ever filled your tank with gravity? Do you have half an hour? Ever lifted 16 gallons of fuel? It's 120 pounds. I suggest you use two hands.

Most of the biofuels movement is not focused on sustaining our relationship to this garden planet. Biodiesel is dominated by companies that are driven by shareholder value, and shareholder value can often equal greed. Shifting our dependence from Exxon to Archer Daniels Midland seems like a very small step toward sustainability to me.

The real question is: "What do we want?" The real answer is that we don't know. What we want in our vehicles is loaded with conflicting desires. When we don't know what we want, we wreak havoc — and that is exactly what we are doing with our vehicle choices today.

If what you want is sustainability, lose your vehicle. Last night, over dinner, someone bemoaned the fact that despite our best efforts we are still plugged into the "transportation part" of our culture, and we discussed how these scenic country lives of ours are wholly unsustainable. The fact that we drive to town to pick up groceries on homemade B100 derived from waste vegetable oil misses the point entirely.

It always struck me as odd that Stacey, who has lived next to the refinery from the beginning and has been supportive of our efforts, never shed her gasoline vehicle to get into a diesel. Now she is shedding her country house and moving to an apartment in town where she can meet her transportation needs on foot. That approach seems much more sustainable than roaring around in our big diesel vehicles.

Some of the experts say that world oil supplies have peaked and we had better come up with some alternatives if we are to sustain our standard of living. Some people think that our standard of living is unsustainable and needs to be abandoned. Some people, like architect Bill McDonough, think that sustainability is too lame a word and that we need to seek abundance, or "fecundity," in our efforts to solve the Earth's environmental problems.

Last year saw about 35 million gallons of biodiesel consumed in this country. Poetic translation: we spilled more crude between 9:00 and 9:15 this morning than we used in biodiesel all last year. Yet more and more people are joining the co-op all the time. More and more people are showing up at our door looking for fuel. And there is some sentiment that the current administration in Washington, who have been referred to as "the coalition of the drilling," will so mismanage our energy policy that we will be compelled to turn to biofuels.

Biodiesel attracts people for a whole bunch of reasons. Some come to it because it is renewable. Some like it for its emissions. Some like it because it can be made in America. Some are drawn to biodiesel so they can avoid taxes. Some so they can be "off grid." And some because they can get subsidies and tax credits. Some even come to biodiesel for profit. And some really do come in the name of world peace, because they feel there is no war required to get the stuff.

Can we agree that we need to replace petroleum? Its pollution affects the health of our children. And when our air is polluted we can't build more smokestacks, and when we can't build more smokestacks we can't sustain our economic development. If we want to build more smokestacks, by golly, we'd better reduce the emissions coming out of our exhaust pipes. Let's walk more. Let's buy hybrids. Let's carpool. Let's make fewer trips. And let's switch to biodiesel.

Biodiesel fits nicely into our current economic model. Growing the feedstocks can help our farmers. Manufacturing biodiesel creates local jobs. And burning it puts less pollution into the air, which leaves us free to pollute somewhere else. When you add the fact that you don't need fighter jets to escort soybeans from Iowa to Virginia, biodiesel looks very good indeed.

It is easy to agree that we want the economic benefits of biodiesel, and it is easy to agree that reducing our dependence on petroleum is a good thing, but we can't agree on how to get the job done. If we had only BTUs to deal with, instead of people and personalities, we would be set.

I'm hoping that petroleum replacement is not as elusive as world peace. It should be easier than that. Surely all the factions of biodiesel should be able to get along. Those who run on straight vegetable oil might stop looking down their noses at the backyard brewers. And those who brew their own might stop looking down their noses at those who run around on "expensive" store-bought biodiesel. Large commercial producers could stop looking down their noses at the small co-ops. Those on B100 could be more welcoming of those (including me) who are trying blends.

I'd like to think that everyone in biodiesel is working hard at petroleum replacement, and that's a good thing for our communities. It's good for our air quality. It's good for our farmers. It's good for our economy and it should be very good for world peace.

Epilogue

A lot has changed on the biodiesel frontier since this book was written. As a chronicle it provides a snapshot of time.

One of the interesting shifts that occurred since this book began is in the policy layer. The one-dollar tax credit subsidy that appeared as part of the Jobs Creation Act in the fall of 2004 has wound its way through the distribution channel in strange and mysterious ways, and now, thanks to the Energy Bill, it will live on until 2008.

The Energy Bill also sets targets for fuel usage in the form of a Renewable Fuel Standard for biodiesel and ethanol — 7.5 billion gallons by 2012. This standard is essentially meaningless drivel that resides in a bill that offers great handouts to all energy sources, regardless of pollution, technology, or cost.

Another fascinating shift has been the market price of a barrel of crude. New record highs have been set month after month, and prices at the pump have risen dramatically.

I sometimes wonder about the fellow who had me abandon my peak oil arguments when he said, "Hell they got oil in Saudi Arabia that they ain't even found yet." I wonder if when he is filling his tank he ever revisits the suggestion that oil has peaked.

The intersection of subsidy and market are providing downward price pressure on biodiesel, and causing demand to skyrocket. Once this renewable fuel is cheaper than fossil, the biodiesel industry will explode into the public consciousness.

George Bush visited Virginia Biodiesel and gave a speech that excited the industry. Many view the dollar tax credit as a good thing. At this point it goes to petroleum blenders and B100 is ineligible, but it does represent a handsome handout to biodiesel.

But despite these efforts, this administration will be remembered by history as the one that was asleep at the wheel when our economy was re-tooling from carbon to renewables. Their understanding of energy has conservation as a virtue rather than imperative. They believe that less regulation and more handouts will increase our energy supply. Both notions are at best risky and at worst fool hardy.

Despite the many convulsions and changes that the biodiesel industry has experienced since this chronicle was set in motion, one thing remains the same: biodiesel is well poised to enter both the vocabulary and the fuel tanks of the driving public.

Bibliography

Alovert, Maria "Mark." *Biodiesel Homebrew Guide.* 10th ed., n.p., 2005. Available from <www.localb100.com/book.html>.

Briggs, Michael. *Widescale Biodiesel Production from Algae* [online]. [Cited April 11, 2005]. UNH Biodiesel Group. <www.unh.edu/p2/biodiesel/article_alge.html>.

Caldara, Anna Maria. *Endangered Environments: Saving the Earth's Vanishing Ecosystems.* Mallard Press, 1991.

Deffeyes, Kenneth S. *Hubbert's Peak: The Impending World Oil Shortage.* Princeton University Press, 2001.

End of Suburbia, The. Gregory Green, 78 min. The Electric Wallpaper Company, 2004. [videocassette].

Estill, Lyle. "Biodiesel: How Grow Your Own has Taken on a Whole New Meaning." *Private Power Magazine.* Spring/Summer 2004, pp. 16-19.

Fat of the Land. Nicole Cousino et al., 55 min. 1995. [videocassette].

Hawken, Paul, Amory Lovins and L. Hunter Lovins. *Natural Capitalism: Creating the Next Industrial Revolution.* Little, Brown and Company, 1999.

Heinberg, Richard. *The Party's Over: Oil, War and the Fate of Industrial Societies.* New Society Publishers, 2003.

Heinberg, Richard. *Powerdown: Options and Actions for a Post-Carbon World.* New Society Publishers, 2004.

Knothe, Gerhard, Jon Van Gerpen and Jürgen Krahl, eds. *The Biodiesel Handbook.* AOCS Press, 2005.

Pahl, Greg. *Biodiesel: Growing a New Energy Economy.* Chelsea Green Publishing Company, 2005.

Plocher, Kumar. *Fueled for Thought* [online]. [Cited April 11, 2005]. <www.livejournal.com/~ybiofuels>.

Prize, The. William Cran, 480 min. Majestic Films International, Public Media Video, 1992. [videocassette].

Rifkin, Jeremy. *The Hydrogen Economy: The Creation of the Worldwide Energy Web and the Redistribution of Power on Earth.* Tarcher/Putnam, 2002.

Schumacher, E.F. *Small is Beautiful: Economics as if People Mattered.* Blond and Briggs, 1973.

Spring, Cari. *When the Light Goes On: Understanding Energy.* Emerald Resource Solutions, 2001.

Tickell, Joshua. *From the Fryer to the Fuel Tank: The Complete Guide to Using Vegetable Oil as an Alternative Fuel.* 3rd ed., Tickell Energy Consultants, 2000.

Tyson, K. Shaine. *2004 Biodiesel Handling and Use Guidelines.* U.S. Department of Energy, 2004

Van Gerpen, Jon et al. *Biodiesel Production Technology.* National Renewable Energy Laboratory, 2004.

www.biodieselamerica.org

New home of Joshua Tickell's Veggie Van.

www.biodiesel.org

Home of the National Biodiesel Board.

www.biofuels.coop

Home of Piedmont Biofuels, and Energy Blog

www.evworld.com   Home of Electric Vehicle World, which covers all sorts of alternative fuels.

www.journeytoforever.org

Keith Addison's site on sustainability of all kinds.

www.localb100.com/book.html

Girl Mark's Biodiesel Homebrew Guide

www.localb100.com/cbt

Home of the Collaborative Biodiesel Tutorial.

www.nrel.gov/homer

An energy balance calculator for all applications.

www.veggieavenger.com

A resource for all things biodiesel.

Index

About the Author

Lyle Estill started making fuel for his tractor in the summer of 2002, and is currently in the midst of opening a large scale biodiesel facility. Prior to being the V.P. of Stuff for Piedmont Biofuels, he has been employed as a furniture restorer, a commercial  beekeeper, a traveling salesman, a computer industry executive, and a metal sculptor.

He lives with his family deep in the woods of Chatham County, North Carolina, where he has been restoring an old farmhouse and home site for the past fifteen years. He has received numerous awards for his work in the arts and on employment issues, including his involvement as an environmental educator.

Sometimes, when the house grows quiet, he sits at the kitchen table and publishes essays in Energy Blog. You can follow his blog at www.biofuels.coop/blog.

If you have enjoyed *Biodiesel Power*, you might also enjoy other

BOOKS TO BUILD A NEW SOCIETY

Our books provide positive solutions for people who
want to make a difference. We specialize in:

**Environment and Justice • Conscientious Commerce
Sustainable Living • Ecological Design and Planning
Natural Building & Appropriate Technology • New Forestry
Educational and Parenting Resources • Nonviolence
Progressive Leadership • Resistance and Community**

New Society Publishers

ENVIRONMENTAL BENEFITS STATEMENT

New Society Publishers has chosen to produce this book on recycled paper made with
100% post consumer waste, processed chlorine free, and old growth free.

For every 5,000 books printed, New Society saves the following resources:[1]

31	Trees
2,833	Pounds of Solid Waste
3,117	Gallons of Water
4,066	Kilowatt Hours of Electricity
5,150	Pounds of Greenhouse Gases
22	Pounds of HAPs, VOCs, and AOX Combined
8	Cubic Yards of Landfill Space

---

[1]Environmental benefits are calculated based on research done by the Environmental Defense Fund and
other members of the Paper Task Force who study the environmental impacts of the paper industry.

For a full list of NSP's titles, please call **1-800-567-6772** *or check out our web site at:*

www.newsociety.com

NEW SOCIETY PUBLISHERS